CALCULATED RISKS

Homeland Security Series

Series Editors:
Tom Payne, University of Southern Mississippi, USA
Tom Lansford, University of Southern Mississippi, USA

This series seeks to provide a body of case studies to explore the growing importance and prominence of homeland security to national defence policy and to examine the development of homeland security within the broader context of national defence policy in the United States and other major developed states. This series will identify and analyze the major threats that are particular to homeland security, as well as those that affect broader national security interests. Comparative studies will be used to elucidate the major similarities and differences in how states approach homeland security and works which advocate new or non-traditional approaches to homeland security. The series aims to integrate information from scholars and practitioners to provide works which will influence the policy debate and examine the ramifications of policy.

Also in the series

The State and Terrorism
National Security and the Mobilization of Power
Joseph H. Campos II
ISBN 978-0-7546-7192-3

Comparative Legal Approaches to Homeland Security and Anti-Terrorism
James Beckman
ISBN 978-0-7546-4651-8

To Protect and Defend
US Homeland Security Policy
Tom Lansford, Robert J. Pauly, Jr and Jack Covarrubias
ISBN 978-0-7546-4505-4

Calculated Risks

Highly Radioactive Waste and Homeland Security

KENNETH A. ROGERS
Coastal Carolina University, USA

MARVIN G. KINGSLEY
Arkansas Tech University, USA

ASHGATE

Published by
Ashgate Publishing Limited
Gower House
Croft Road
Aldershot
Hampshire GU11 3HR
England

Ashgate Publishing Company
Suite 420
101 Cherry Street
Burlington, VT 05401-4405
USA

Ashgate website: http://www.ashgate.com

British Library Cataloguing in Publication Data
Rogers, Kenneth A., 1946-
   Calculated risks : highly radioactive waste and homeland
   security. - (Homeland security)
   1. Radioactive waste disposal - Risk assessment - United
   States 2. Radioactive waste disposal - Government policy -
   United States 3. Radioactive waste disposal - Management -
   United States
   I. Title II. Kingsley, Marvin G.
   363.7'289'0973

Library of Congress Cataloging-in-Publication Data
Rogers, Kenneth A., 1946-
   Calculated risks : highly radioactive waste and homeland security / by Kenneth A.
Rogers and Marvin G. Kingsley.
      p. cm. -- (Homeland security)
   Includes bibliographical references and index.
   ISBN 978-07546-7133-6
   1. Radioactive waste disposal--Risk assessment--United States. 2. Radioactive waste
disposal--Government policy--United States. 3. Radioactive wastes--Management--United
States. I. Kingsley, Marvin G. II. Title.

   TD898.14.R57R65 2007
   363.72'89--dc22
                                                                        2007014535

ISBN 978-0-7546-7133-6

Printed and bound in Great Britain by MPG Books Ltd, Bodmin, Cornwall.

Contents

List of Figures

List of Tables

Preface

Since the dawn of the atomic age in 1942 with the first controlled, self-sustaining chain reaction of radioactive material, there has been considerable controversy over the desirability of splitting the atom. On the one hand, nuclear energy unleashed the creation of a terrible weapon of mass destruction. On the other hand, the promise of an endless supply of a relatively affordable and reliable source of electricity gave the promise of a new era in energy production. While commercial nuclear power has proved to be a reliable source of electricity that does not create air pollution or contribute to global warming, it does generate highly radioactive waste that will be lethal for thousands of years.

While the problem of nuclear waste was first recognized in the early 1950s, it was generally accepted that it was not a serious concern at the time, and any decision on what to do with the waste could be postponed to a later date. Over 50 years have passed since then and policy makers are still struggling to develop a consensus on how best to approach the problem of highly radioactive waste storage and disposal. The inability of decision makers to come to grips with the problem of what to do with the highly radioactive waste has delayed the development of a rational policy for coping effectively with the ever-increasing amounts of radioactive waste being generated. This delay basically has meant that spent nuclear fuel generated at commercial nuclear power facilities and high-level waste produced at Department of Energy defense facilities continue to be stored on-site since neither a centralized interim storage option nor a permanent disposal solution has been found. This delay primarily has been due to the conflict generated by the politicization of the issue.

This book is a case study of highly radioactive waste storage and disposal policy. The basic premise of this study is that in spite of the pressing homeland security and energy security concerns associated with highly radioactive waste being stored on-site at numerous locations around the country, political considerations have prevented policymakers from adopting a long-term solution to the problem.

Chapter 1 discusses the homeland security, energy security and societal implications of highly radioactive waste storage and disposal policy. It also examines the impact of the structure of the political system (e.g., federalism and separation of powers) and the myriad of actors involved in radioactive waste policymaking. Chapter 2 looks at on-site radioactive waste storage. It examines the history of on-site storage, the types of on-site storage, and the advantages and disadvantages associated with each type of storage option. Chapter 3 provides an overview of centralized interim radioactive waste storage and focuses on two specific projects – Skull Valley (Utah) and Owl Creek (Wyoming). Although not as contentious as the Yucca Mountain project, interim storage proposals also have generated considerable opposition. Chapter 4 addresses the long-term disposal of highly radioactive waste and focuses on the Yucca Mountain project. The Yucca Mountain initiative has become heavily politicized due to the perceived unfairness of the site selection process as well as apprehension over the geologic and environmental suitability project site. Chapter 5 analyzes transportation concerns associated with highly radioactive waste such as security and safety of the nuclear material during transport, and emergency response capabilities in the event of an incident. Chapter 6 provides a conceptual framework for what should, and ultimately what does, drive highly radioactive waste policymaking. It examines not only the homeland security and energy security concerns associated with highly radioactive waste storage and disposal, but also the political ramifications of various storage and disposal options. Chapter 7 provides an overview of the factors affecting U.S. radioactive waste policy; a discussion of what to do, some recommendations for storage, disposal, and transportation policy, and some concluding thoughts on the future of radioactive waste policymaking.

One of the strengths of this book is the extensive research that has gone into the presentation of the issues. This provides the reader with a substantial list of up-to-date and relevant sources for each topic addressed. Moreover, there has been a focus on the use of web sources, which provide the reader with an easy access to a variety of sources readily accessible on the Internet. However, it should be noted that while every attempt has been made to ensure the currency of the website address, they can and do change frequently. In addition, a list of terms has been provided at the end of each chapter to highlight key terms discussed in the readings. For ease of understanding the definition of these key terms, a glossary has been included which provides a brief definition for each term.

In the final analysis, the problem of what to do with the ever-increasing amounts of highly radioactive waste is a pressing issue that must be dealt with in the near-term to avoid a real and visible threat to the country's security. The homeland security and energy security concerns are too great to ignore. Ultimately, for the good of the country, political considerations must be set aside in order to move forward and develop a rational and coherent national radioactive waste storage/disposal policy.

Kenneth A. Rogers
Marvin G. Kingsley

Acknowledgments

As is the case with any endeavor of this magnitude, others have contributed immeasurably to its success. We are especially grateful for the support offered by our families whose encouragement and understanding was essential in being able to complete this project. We would like to extend a special appreciation to Donna Rogers for her technical support throughout the process, especially in coping with the computer hardware and software problems. Her help was critical to completing the manuscript. Also, it was her initial research on the Yucca Mountain Project that sparked an interest in the topic of radioactive waste.

We also wish to express our gratitude to Arkansas Tech University for its support of both the research and writing of the manuscript. The research assistance provided through the Arkansas Tech Undergraduate Research Program was instrumental in getting this project off the ground, and Tech's support for a sabbatical allowed the project to be completed in a more timely fashion. We also would like to thank the staff of the Ross Pendergraft Information and Technology Center located on the campus of Arkansas Tech University for their research assistance in locating and obtaining sources and guidance on utilizing their state-of-the-art research database platforms and U.S. Government Documents Repository. Finally, we would like to thank the Arkansas Center for Energy, Natural Resources and Environmental Studies for its support for the research on interim radioactive waste storage.

We also would like to acknowledge the contributions of a number of individuals who were involved in the various stages of completing the manuscript. First, we would like to extend a special thanks to the staff at Ashgate Publishing for their expert advice and assistance, especially Margaret Younger who was very responsive and supportive throughout the editorial process. Thanks to Wilma LaBahn, Angela Rogers and Harvey Young for their insightful comments which greatly aided editing of the manuscript. Their support, helpful criticisms, and suggestions were instrumental in improving the quality of the writing. We also would like to thank Dr. John Navin for his suggested title *Calculated Risks*. We also would like to express our gratitude to the International Association of Emergency Managers for lending an opportunity to showcase elements of the research to emergency management industry professionals at their annual conference in Columbus, Ohio during the Fall of 2002. Finally, we would like to thank Sue Martin, spokesperson for Private Fuel Storage, for her responsiveness and willingness to share her perspectives concerning the interim radioactive waste storage initiative at Skull Valley, Utah.

List of Abbreviations

Atomic Energy Commission (AEC)
Bureau of Indian Affairs (BIA)
Bureau of Land Management (BLM)
Central Interim Storage Facility (CISF)
Department of Energy (DOE)
Department of Transportation (DOT)
Design based threat (DBT)
Energy Information Administration (EIA)
Environmental Protection Agency (EPA)
Federal Emergency Management Agency (FEMA)
Federal Motor Carrier Safety Administration (FMCSA)
Federal Radiological Emergency Response Plan (FRERP)
Federal Railroad Administration (FRA)
Fiscal year (FY)
Government Accountability Office (GAO)
Monitored Retrievable Storage (MRS)
National Academy of Sciences (NAS)
National Association of Regulatory Utility Commissioners (NARUC)
National Transportation Program (NTP)
Not in my backyard (NIMBY)
Nuclear Energy Institute (NEI)
Nuclear Regulatory Commission (NRC)
Nuclear Waste Policy Act of 1982 (NWPA)
Nuclear Waste Policy Amendments Act of 1987 (NWPAA)
Nuclear Waste Technical Review Board (NWTRB)
Office of Civilian Radioactive Waste Management (OCRWM)
Pipeline and Hazardous Materials Safety Administration (PHMSA)
Private Fuel Storage (PFS)
Project on Government Oversight (POGO)
Radioactive Materials Incident Report (RMIR)
Radiological Assistance Program (RAP)
Regional Coordinating Offices (RCO)
Retrievable Surface Storage Facility (RSSF)
Transportation Emergency Preparedness Program (TEPP)
Transportation Resource Exchange Center (T-REX)
Waste Isolation Pilot Plant (WIPP)
Western Governors' Association (WGA)

Chapter 1

Understanding the Problem

The problem of highly radioactive waste storage or disposal now has become one of the most controversial aspects of nuclear technology. As the inventories of spent nuclear fuel from commercial nuclear reactors and high-level waste from defense-related processing plants have continued to mount, the issue has become increasingly contentious and politicized. Paradoxically, as the need for action has become more acute, conflict generated by the politicization of the issue has delayed progress on developing a long-term solution to the problem.

The inability of the U.S. government to meet the 1998 deadline of accepting the commercial spent nuclear fuel and high-level defense waste that is stored on-site around the country has created homeland security and energy security concerns. At the same time, the structure of the American political system ensures that policymaking for contentious issues such as highly radioactive waste storage/disposal will be slow and difficult. This, in turn, has contributed to the delay in achieving a national consensus on how best to approach the radioactive waste storage/disposal issue.

This chapter provides an overview of the highly radioactive waste storage and disposal issue. First, the problems associated with the growing stockpiles of spent nuclear fuel from commercial nuclear reactors and high-level wastes from defense facilities are presented. Homeland security and energy security concerns, as well as the moral and ethical consequences associated with the delay in forging a long-term consensus on how best to cope with the highly radioactive waste issue are briefly discussed. Second, the impact of the structure of the U.S. political system guarantees that a myriad of actors will be involved in the formulation and implementation of highly radioactive waste storage and disposal policy. The interaction of the various federal, state, tribal and local government authorities, as well as the involvement of a multitude of special interest groups and the media, is discussed. These factors ensure that it will be more difficult to arrive at a near-term consensus on how best to address the problem of the growing inventories of highly radioactive waste.

The Problem

The problem of *storage* (the temporary placement) or *disposal* (the permanent placement) of radioactive waste generated minimal concern in the early years after the discovery of nuclear fission since it was not perceived by policy makers to be an issue that required immediate attention. Rather, the disposal of nuclear waste generally was considered to be a problem that could be addressed sometime in the future. This mindset delayed recognition of an impending problem. This, in turn, prevented any real progress on developing a long-term solution to the problem of the mounting inventories of radioactive waste, and essentially allowed policy makers to avoid the issue until it became a more pressing and immediate problem. Even after it became evident that the disposal problem of highly radioactive waste demanded more immediate action, effective policies to develop a long-term solution to the problem have been rather slow to materialize. As one observer of the process of coping with highly radioactive waste has pointed out, "The scientific problem of finding a geologically suitable site would be much easier than the political problem of finding a state willing to take the waste." (McCutcheon, 2002: 88)

As early as 1957, the National Academy of Sciences (NAS) recommended the long-term disposal of radioactive waste underground in salt beds and salt domes as the most practical solution to the problem. (NAS, 1957: 1) However, it was not until 25 years later that the federal *Nuclear Waste Policy Act of 1982 (NWPA)* began the process of establishing a deep, underground permanent national repository for highly radioactive waste. The 1982 Act mandated that the Department of Energy (DOE) would study multiple sites in the West and select one to be ready to serve as the first national repository by 31 January 1998. A second site in the East, where the majority of highly radioactive material is generated, would be identified later. The site selection process was slow and arduous due to opposition caused by states identified as potential hosts for the facility. Eventually, under pressure from congressional members from the states targeted for a possible repository, the Department of Energy, and the nuclear power utilities lobby, Congress passed the *Nuclear Waste Policy Amendments Act of 1987 (NWPAA)* which amended the original Act and singled-out Yucca Mountain, Nevada, as the only site to be studied. The perceived unfairness of the site selection process, coupled with geologic and environmental uncertainties of the host site, has caused considerable political conflict at the national level.

At the same time, efforts to construct an interim storage facility also have been fraught with conflict. Clearly, the interim storage issue is linked to progress on a permanent repository since opponents of an interim nuclear waste storage facility have voiced concern that any interim facility could become a de facto repository for highly radioactive waste if a permanent disposal option is not found. They argue that once an interim facility is established, political realities might preclude building a permanent repository. Moreover, opponents also have expressed concern over the safety of transporting large amounts of highly radioactive waste over long distances – sometimes through heavily populated areas. The lack of resolution of where to store or dispose of highly radioactive waste has forced operators of commercial nuclear power reactors and DOE facilities to store both spent nuclear fuel and high-level radioactive waste on-site at numerous plants and facilities across the country.

Highly radioactive waste has continued to pile up at reactors across the country at 125 sites in 39 states (see Figure 1.1 – Locations Where Highly Radioactive Waste is Stored). Today, over half of the U.S. population lives within a few miles of one of these sites. According to the Office of Civilian Radioactive Waste Management (OCRWM), commercial nuclear power reactors have generated over 50,000 metric tons of spent nuclear fuel. (OCRWM, 2004a: 4) At the same time, high-level waste from defense facilities, nuclear naval ships, and surplus plutonium from nuclear weapons production, has generated significant quantities of high-level nuclear waste. In spite of the fact that highly radioactive waste continues to accumulate at sites across the country, there has been a surprising lack of consensus on how best to address the problem.

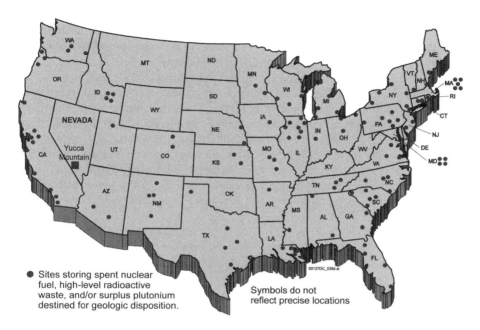

● Sites storing spent nuclear fuel, high-level radioactive waste, and/or surplus plutonium destined for geologic disposition.

Symbols do not reflect precise locations

Figure 1.1 Locations Where Highly Radioactive Waste is Stored
Source: "Current Locations of Spent Nuclear Fuel (SNF) and High-Level Radioactive Waste (HLW)," OCRWM Program Briefing, Office of Civilian Radioactive Waste Management, U.S. Department of Energy (OCRWM, 2004a: 5).

Homeland Security and Energy Security Concerns

The inability of U.S. policymakers to come to grips with the problem of the continued generation of radioactive waste has both *homeland security* (keeping highly radioactive waste secure from terrorism) and *energy security* (not being forced to shut down commercial nuclear power facilities due to a lack of sufficient on-site storage space) implications. The events of 9/11 have focused attention on the vulnerability of nuclear power facilities. At the same time, the lack of sufficient on-site storage space for spent nuclear fuel threatens the continued electric power generation capacity at a number of commercial nuclear power facilities since some of the plants could be forced to shut down prematurely due to insufficient on-site storage capability.

Homeland Security Concerns Storing large quantities of highly radioactive waste at numerous locations across the country heightens anxiety over a possible terrorist attack with the potential for a significant release of radiation. Over the years, numerous threats have been directed at commercial nuclear power facilities. Concern over security has increased since the terrorist attacks on 9/11 – especially since nuclear power facilities have been singled out as a prime target for a future attack. In 2002, the Director of the U.S. Nuclear Regulatory Commission (NRC) acknowledged in a report to Congress that security "weaknesses" were identified in almost half of the mock security exercises conducted at nuclear power plants during the 1990s. (Meserve, 2002: 27)

While much of this concern over security has been directed towards an attack on the reactors themselves, they are less of a problem since they are housed within steel and concrete reinforced containment buildings. However, the highly radioactive waste stored on-site is more vulnerable since the spent nuclear fuel is stored either in cooling pools within non-reinforced buildings or stored in dry casks above ground. Thus, the spent fuel is more worrisome since it would be more vulnerable to any terrorist attack. In fact, in 2002, Secretary of Energy Spencer Abraham cited concerns over maintaining large amounts of highly radioactive waste across the country within miles of millions of Americans as a threat to homeland security. (Abraham, 2002: 3)

Energy Security Concerns Concerns over adequate energy supplies first surfaced in the United States during the Arab oil embargo in the early 1970s. Oil shocks in subsequent years and the recent dramatic rise in oil prices have reinforced the perception of the need for the United States to reduce its dependency on foreign sources of energy. However, this can only be done if the United States does not lose any of its current domestic energy generation capacity and is able to keep pace with future energy demands. Since nuclear power generated electricity accounts for about one-fifth of the total U.S. electric power production, any forced premature closure of nuclear power facilities will have a significant negative impact on energy security by increasing U.S. reliance on foreign sources of oil and natural gas. Moreover, according to the U.S. Energy Information Administration (EIA), U.S. electricity generation needs are projected to increase by about 50 percent by 2025. (EIA, 2004: 145)

Due to the anticipated demand for new energy sources, in 2001 the Bush Administration called for the expansion of nuclear generated electricity as part of its national energy policy.[1] Four years later, Congress passed The Energy Policy Act of 2005 (P.L. 109-58, 2005) which provides incentives to the nuclear power industry to build new generation reactors by streamlining the application process, providing guarantees and extending the Price-Anderson Act to 2025 for new nuclear

---

1   See (White House, 2001) "National Energy Policy" for a discussion of the Bush Administration energy plan. [http://www.whitehouse.gov/energy/Forward.pdf] Also, see "DOE Researchers Demonstrate Feasibility of Efficient Hydrogen Production from Nuclear Energy," Press Release, November 30, 2004 [http://www.doe.gov/news/1545.htm] for a discussion of using new generation nuclear reactor technologies to produce hydrogen for powering future vehicle fuel cell technology (DOE, 2004b: 1).

power reactors.[2] In support of the new energy plan, DOE has developed the Nuclear Power 2010 program, which would facilitate the licensing process for building new nuclear power plants. (DOE, 2004c: 1) At the same time, the commercial nuclear power industry has developed the Vision 2020 plan to increase nuclear electricity generation by 50,000 megawatts from *new* nuclear power plant capacity by 2020. (NEI, 2004: 1) However, it will be difficult to maintain current electric generating capacity, let alone expand production, if some commercial nuclear power facilities are forced to shut down prematurely due to a lack of sufficient on-site nuclear waste storage space. Thus, the problem of continued highly radioactive waste generation has both homeland security and energy security implications.

Societal and Political Implications

The delay in forging a coherent and rational public policy on highly radioactive waste also has important societal implications. Paradoxically, as the need for action has become more acute, the conflict generated by the politicization of the issue has delayed progress on developing a viable long-term solution to the problem. The policy process has been quite contentious for a number of reasons:

> It [radioactive waste disposal] is a complicated and distant threat, of little interest to the public or to most policymakers. It has few immediate payoffs and many immediate costs. It is controversial, and most of the solutions put forth for it are controversial, unlikely to appeal to a majority of interests. (Katz, 2001: 23)

Conflict associated with the current on-site storage regime has been relatively low. The concept of centralized interim highly radioactive waste storage has been controversial since opponents have voiced concern that any interim facility could become a de facto repository. Thus, any interim storage solution has been inextricably tied to progress on long-term disposal. The long-term disposal of highly radioactive waste has been the most politically contentious option primarily because the "not in my backyard" (NIMBY) reaction has generated opposition in states designated as a potential host for a nuclear waste repository.[3] Moreover, resentment over the perceived unfairness of the site selection process coupled with apprehension over the geologic suitability of the Yucca Mountain site has generated additional controversy, which has intensified the conflict. As a result, the attempt to construct a geologic repository at Yucca Mountain has become intensely politicized.

Because of the conflict associated with the proposed centralized interim storage and permanent disposal initiatives, it has been easier for policymakers to continue the relatively non-controversial status quo policy (especially one that has been in place

---

2   See (PL 109-58, 2005) "Energy Policy Act of 2005," Public Law 109-58, August 8, 2005 [http://frwebgate.access.gpo.gov/cgi-bin/getdoc.cgi?dbname=109_cong_public_laws&docid=f:publ058.109.pdf]

3   For an excellent discussion on the NIMBY phenomenon, see (Munton, 1996) Munton, Don. "Introduction: The NIMBY Phenomenon and Approaches to Facility Siting," in *Hazardous Waste Siting and Democratic Choice: The NIMBY Phenomenon and Approaches to Facility Siting*, Georgetown University Press, 1996.

for a number of years) of on-site storage at the point of generation. Policymakers generally prefer to follow existing policy, or make only minor policy adjustments (i.e., incremental policymaking), since it involves minimal short-term political risk. At the same time, elected officials tend to focus on the short-term – the next election.

Moral and Ethical Implications

The delay in forging a coherent and rational public policy on highly radioactive waste also has moral and ethical consequences. One of the major determinants of policy in a democracy is the distribution of the perceived costs and benefits of a particular course of action. So, what are the costs and benefits associated with highly radioactive waste policy? Basically, the type of storage or disposal regime will determine the costs. For example, on-site storage at the point of origin distributes costs to the communities who also reap the benefits of relative clean and cost-effective electricity production, increased tax base, and the economic benefit of having a well-educated, well-paid workforce. On the other hand, any centralized interim storage and permanent disposal option will concentrate the costs (safety and health) to the communities who host the storage and disposal facility.

Thus, it will be important not only to ensure that any host community receives adequate compensation for the potential safety and health risks incurred as a result of hosting the storage or disposal site, but also ensure that safety is maximized so that the host community does not incur a disproportionate share of the burden in terms of risk. The question of equity also has been a recurring issue with respect to the site selection process for both centralized interim storage and permanent disposal. For example, while most of the spent nuclear fuel from commercial nuclear power facilities is generated in eastern states, only western states (New Mexico, Nevada and Utah) have been selected to host either a large-scale centralized interim storage or permanent disposal facility.

Another equity concern is who pays for the national radioactive waste repository. The 1982 NWPA established the Nuclear Waste Fund to provide a source of funding for the national repository. In spite of the fact that the commercial nuclear power utilities and ratepayers have paid billions of dollars into the Fund, they still are burdened with the costs of storing the waste on-site. At the same time, monies from the Fund have been diverted to other government programs, which has made it difficult to fully fund the Yucca Mountain project.

Finally, it is important to ensure that future generations are not saddled with the burden of what to do with the nuclear waste. The concept of *rolling stewardship*, where the risk of managing the radioactive waste will "roll over" from generation to generation, does not adequately address equity concerns. (Falcone and Orossco, 1998: 760-763) Thus, the generation that creates the radioactive waste should be the same generation that is responsible for its proper disposal. It is imperative that some equitable resolution of the problem be attained in the near-term so that future generations do not bear the safety, health and economic costs of disposing of the highly radioactive waste.

Impact of the Political System

The American political system was designed by the Founders to ensure that no single entity would be able to accumulate too much power at the expense of the people. The principles of limiting the power of government (e.g., guaranteeing certain basic rights and liberties) and diffusing the power granted to government (e.g., federalism and the separation of powers) are a foundation of the American political system. While this structure has prevented an abuse of power by government, it also has ensured that policymaking – especially for contentious issues such as highly radioactive waste – would be slow and arduous. Thus, policy changes likely will be incremental in nature (i.e., only slow, gradual changes in policy). Generally, only when a problem is perceived to be a crisis is it possible to have a significant shift in policy direction. Since highly radioactive waste historically has not been perceived to be an issue that needed to be addressed in the near-term, there has been little incentive for policymakers to arrive at a final solution.

Over the years, the U.S. political system has evolved to a point where there are many avenues available to influence policy. Thus, the debate on the more contentious issues such as highly radioactive waste can be lively and strident. The structure of the U.S. political system (e.g., federalism and the separation of powers) ensures that a multitude of actors will be involved in policymaking and that the process will be protracted and difficult.

Federalism

Federalism initially was established to ensure a diffusion of power away from the national government. A two-tiered system of government was created where the central (federal) government would be sovereign in its sphere of power, and the regional (state) governments would be sovereign in their respective spheres of power. As the twentieth century progressed, however, the federal government began to accumulate more power at the expense of the states. The principle of national supremacy, the interstate commerce clause, and the superior financial resources of the federal government have been the primary vehicles used to increase federal power.

The principle of *national supremacy* is enshrined in the Constitution (Article VI) where federal laws "shall be the supreme law of the land." Thus, if a federal law and state law conflict, the federal law takes precedence over the state law. Thus, any attempts by states to regulate radioactive waste storage/disposal must not conflict with federal regulations. At the same time, the federal courts have used the *interstate commerce clause* in the Constitution (Article I, Section 8), which gives the federal government control over activities that transcend state boundaries, to extend federal control over a number of areas that previously were under control of the states. Since any centralized interim storage and permanent disposal solution will necessitate transporting significant quantities of nuclear waste across states, federal laws will take precedence. Finally, the superior financial resources of the federal government ensures that there will be a degree of control over state activities since the states have become dependent on federal monies to operate many programs. Thus, the federal government can dictate policy by offering financial incentives, or withholding monies if a state does not adhere to federal guidelines. Consequently, with federalism today the preponderance of power has shifted to the federal government.

Federalism has ensured a certain amount of conflict between the two levels of government. In essence, the states generally do not want to host a centralized interim storage facility or a permanent repository due to the political fallout that such a facility would generate. On the other hand, the federal government understands that the homeland security and energy concerns associated with having highly radioactive waste stored at numerous locations across the country dictates that some type of centralized solution is necessary. Thus, the stage is set for significant conflict between the federal government and the states identified to host a centralized highly radioactive waste storage or disposal facility.

Separation of Powers

The separation of powers between the legislative, executive, and judicial branches creates a system of checks and balances. This ensures that devising any radioactive waste policy will be slow and arduous since all three branches have to concur on policy. While the president can propose policy such as recommending Yucca Mountain for the national repository, it is Congress that has the ultimate responsibility for passing the legislation to establish the facility. Moreover, Congress appropriates the money to fund construction and operation of any federally sponsored centralized interim storage or permanent disposal facility. While the president can threaten to veto, or even, in fact, veto radioactive waste legislation as President Clinton did in the 1990s, Congress has the ability to override the veto with a two-thirds majority in both chambers. However, the political realities are such that it is difficult for Congress to override a presidential veto. For example, since the founding of the Republic fewer than ten percent of presidential vetoes have been successfully overridden. Even if the executive and legislative branches agree on a particular course of action, the judicial branch can block action on any radioactive waste initiative by virtue of exercising judicial review which allows the courts to declare executive and legislative acts as unconstitutional and thus, null and void. Since highly radioactive waste litigation has increased substantially in recent years, the courts are now more active in the radioactive waste policy debate.

The Actors

The United States can be characterized as a pluralist democracy where competing groups and shifting alliances help to determine public policy. As a result, a myriad of actors are involved in the radioactive waste debate which ensures that politics will play a major role in the formulation and implementation of highly radioactive waste policy. Government actors include the entire gamut of federal, state, tribal, and local authorities. Interest groups have been especially active; nuclear energy utilities are in strong support of establishing a centralized storage or disposal facility, while environmental and some citizen activist groups have been vocal in their opposition to such a facility. The media can play an important role in shaping citizen perceptions as they report on the various radioactive waste issues.

Federal Government

There are a number of important federal actors involved in formulating and implementing highly radioactive waste policy. Clearly, the president sets the tone by which policies are proposed, Congress ultimately establishes the policy by passing legislation, the bureaucracy plays a major role by implementing policy, and the courts establish policy through their rulings.

White House The president sets the tone and direction of nuclear waste policy by virtue of the policies pursued and the appointments made for key positions such as the Secretary of Energy. For example, President Clinton moved cautiously on the Yucca Mountain initiative due primarily to the mounting political opposition. In the 1990s, the administration either threatened to veto, or vetoed, several congressional legislative initiatives to move forward on the Yucca Mountain project and interim storage proposals. Because congressional Democrats were reluctant to challenge the White House and further damage the president's political power (already compromised by the independent prosecutor probes and subsequent impeachment proceedings), there was insufficient political support for the Yucca Mountain project. In contrast, the Bush Administration exhibited a substantial amount of political support for the Yucca Mountain initiative during the first term, formally recommending its approval to Congress in 2002. The Secretary of Energy, Spencer Abraham, was a forceful and articulate spokesman for the Yucca Mountain project. In his letter to the president recommending approval of the Yucca Mountain repository, Abraham cited homeland security concerns and energy security concerns as compelling reasons to move forward on the Yucca Mountain project. (Abraham, 2002: 1)

Political realities, however, dictate that the White House must make difficult choices on what priorities to set since there is only so much political capital that a president has to spend. Thus, it will be interesting to see how aggressive the Bush Administration will pursue the Yucca Mountain initiative during its second term. It is likely that other national security concerns (e.g., Iraq, Afghanistan, and the war on terrorism) and domestic priorities (e.g., aftermath of Hurricane Katrina, Social Security reform, and tax reform) will deflect the Administration's attention away from aggressively pursuing a national repository for highly radioactive waste. The question becomes, will the Administration lower its priority for the Yucca Mountain project in order to gain political support for other initiatives, or will it press forward aggressively to address the perceived security and energy concerns? It appears that Yucca Mountain already has assumed a lower status in the Administration's domestic priorities since the 2006 fiscal year (FY) budget submitted by the president to Congress called for over a 25 percent cut in Yucca Mountain funding (e.g., $880 million was proposed for the FY 2005, while only $651 million was proposed for the FY 2006 budget). (OMB, 2004: 122; OMB, 2005: 113) For FY 2008, the White House only proposed $494.5 million for Yucca Mountain. (GPO, 2007: 59) Since budgets submitted by the president are regarded as a policy document that reflects administration priorities, it appears that the Yucca Mountain project already has been given a lower priority.

Congress   Congress also has played a major role in highly radioactive waste policy since it passes the legislation that establishes the rules governing the disposition of radioactive waste. In exercising its "responsibility" Congress decidedly targeted one state, to the benefit of others, with the 1987 amendments to the 1982 Nuclear Waste Policy Act. In what has been characterized as "the most politically explosive move in the history of nuclear waste policy," Senator Bennett Johnston (D-LA) proposed legislation amending the 1982 Nuclear Waste Policy Act that specifically singled out Nevada as the host for a permanent national repository for highly radioactive waste. (McCutcheon, 2002: 88) It should be noted that Louisiana was one of the initial six states identified as a potential host for a national highly radioactive waste repository. Thus, Senator Johnston's proposal to designate Yucca Mountain (Nevada) as the only site to be studied deflected attention away from Louisiana as well as the other four states initially identified as potential repository hosts (Mississippi, Texas, Utah, and Washington). Nevada was not able to derail the proposal since it had a small and relatively junior congressional delegation. Congress's adoption of the Johnston proposal has been characterized as "decidedly undemocratic...especially when it came to states with relatively little political clout." (McCutcheon, 2002: 88)

At the same time, Senator Harry Reid (D, NV) – an outspoken critic of the Yucca Mountain project – has had success over the years in reducing the budget for the Yucca Mountain repository. For example, President Bush asked for $591 million for the Yucca Mountain project in the FY 2003 budget. (OCRWM, 2005a) However, as ranking member (i.e., senior senator from the minority party) of the Senate Appropriations Committee at the time, Senator Reid was able to cut funding for Yucca Mountain by over $250 million for the Senate version of the FY 2003 proposed budget. (Abrahams, 2003: 1) While Congress eventually appropriated $457 million in the FY 2003 budget for the Yucca Mountain project, the $134 million cut still represented a 22 percent reduction in funding. (OCRWM, 2005a) Congress has continued to cut funding for Yucca Mountain since, including a 35 percent reduction in FY 2005. (OMB, 2004, OCRWM, 2005b) The Director of the Office of Civilian Radioactive Waste Management stated in 2003 that the target date 2010 for the Yucca Mountain repository was in jeopardy due to the continued budget cuts. (Nuke-Energy, 2002b: 1) In 2005, the Director announced that the 2010 date could not be met.

Further complicating the issue for the Bush Administration was Senator Reid's elevation to minority leader after the 2004 elections. In spite of the fact that the Administration had a friendly majority in both chambers of Congress, it was difficult to push through legislation that was opposed by Democrats since Senate rules allow for the minority party to block legislative initiatives unless there is a 60-vote majority in support of the legislation. Thus, unless the Administration had sufficient support from Senate Democrats, it was difficult to push through any legislative initiative on Yucca Mountain.

Senator Reid's elevation to the position of Senate majority leader after the 2006 congressional elections will make it even more difficult for the Bush Administration to follow through with the Yucca Mountain initiative – especially since Reid has vowed, "Yucca Mountain will not come to be." (Tetreault, 2003: 2) As the leader of the Senate, the majority leader plays an important role in the overall legislative process since he is responsible for scheduling legislation in the Senate. Ultimately, the White House will be forced to deal with Senator Reid on a number of issues, and it is unlikely that the senator will be willing to accommodate any policy that results in the construction of a highly radioactive waste disposal facility in Nevada. Thus, it is possible that the Administration may be forced to lower the priority of the Yucca Mountain initiative (e.g., cut funding) in order to gain support from congressional Democrats for what is deemed to be more pressing issues.

The Bureaucracy A number of bureaucratic agencies have the responsibility and jurisdiction over highly radioactive waste issues. While a variety of government agencies are involved in some aspect of nuclear regulation, there are three primary federal organizations tasked with managing spent nuclear fuel and high-level waste: the U.S. Department of Energy, the U.S. Nuclear Regulatory Commission, and the U.S. Environmental Protection Agency. Additional federal actors also play an important role in highly radioactive waste policy. For example, the U.S. Department of Transportation is involved in radioactive waste transportation issues, the Federal Emergency Management Agency will play a role in the response and recovery phase of a nuclear waste incident, and the Nuclear Waste Technical Review Board (NWTRB) is an independent agency of the U.S. government tasked to provide independent oversight of the U.S. civilian radioactive waste management program.[4]

The *Department of Energy (DOE)* is an important bureaucratic actor and plays a significant role in highly radioactive waste policy in a number of ways.[5] First, DOE is responsible for promoting a diverse supply of reliable, economical and environmentally sound energy – including nuclear energy. Second, the Department has been tasked with the clean up of high-level waste at DOE facilities as a result of the nation's nuclear weapons and research programs. Third, DOE also is tasked to bring together a wide variety of organizations that have an interest in the transportation of radioactive waste materials through its National Transportation Program (NTP). The NTP mission is to provide policy guidance, technical and management support, and operational services to assure the availability of safe and secure transportation of non-classified DOE nuclear materials. (NTP, 2005: 1) Fourth, the Transportation External Coordination Working Group (TEC/WG) brings together a diverse group of interested stakeholders to discuss a wide variety of radioactive waste transportation issues.[6] Finally, DOE is responsible for developing a permanent disposal capacity

---

4   See the NWTRB website [http://www.nwtrb.gov] for more information on its activities monitoring radioactive waste disposal.

5   See the DOE website ("Energy Sources" and "Safety and Security") for an overview of the department's role in radioactive waste policy. [http://www.doe.gov]

6   See (TEC/WG, 2002) "Charter," National Transportation Program, U.S. Department of Energy, Revised 2002 [http://www.tecworkinggroup.org/tecchart.pdf] for a discussion of the responsibilities and objectives of the TEC/WG. Also see (TEC/WG, 2007) "TEC Members," Transportation External Working Group, U.S. Department of Energy [http://www. tecworkinggroup.org/members.html] for a list of the TEC membership.

(e.g., Yucca Mountain) to manage the high-level radioactive waste produced from past nuclear weapons and research programs, as well as spent nuclear fuel generated by commercial nuclear reactors. (DOE, 2004a: 1) The *Office of Civilian Radioactive Waste Management (OCRWM)*, established as a DOE program in 1982 by the Nuclear Waste Policy Act, has been tasked to develop and manage the federal highly radioactive waste system. The OCRWM has the primary responsibility for determining the suitability of the Yucca Mountain project.[7] Once it is determined that the repository should be constructed, DOE will submit a license application to the Nuclear Regulatory Commission for technical review. (OCRWM, 2004b: 1)

The *Nuclear Regulatory Commission (NRC)* is an independent agency established by the Energy Reorganization Act of 1974 to regulate certain nuclear materials, including nuclear waste, in order to ensure the protection of public health and safety from radiation exposure.[8] The NRC is responsible for regulating reactors, nuclear materials and nuclear waste through licensing requirements. (NRC, 2005e: 1) The Office of Nuclear Reactor Regulation has overall responsibility for regulating reactors (NRC, 2005c: 1), and the Office of Nuclear Material Safety and Safeguards (NRC, 2005b: 1; NRC, 2005d: 1) has overall responsibility for the NRC's nuclear materials program and for the radioactive waste regulation program. The Office of Nuclear Security and Incident Response is responsible for developing emergency preparedness programs to respond to a variety of nuclear emergencies including terrorism. (NRC, 2005a: 1) The NRC has overall responsibility for developing regulations to implement safety standards for licensing a national highly radioactive waste repository. Before a highly radioactive waste repository can be constructed, the NRC must follow a two-step licensing process. The first step is to issue a license authorizing construction of the repository. Prior to granting the license, the Commission's Atomic Safety and Licensing Board has the responsibility for conducting public hearings to address concerns with any proposed repository license application. Then, after outstanding issues have been addressed, the NRC must follow-up by issuing a license to receive and possess the waste before the repository's operations can begin. (OCRWM, 2004c: 1-2) The NRC also is responsible for regulating the performance of packaging and transport operations of shippers of spent nuclear fuel and high-level radioactive waste. (DOE, 2002: 4)

The *Environmental Protection Agency (EPA)*, through its Radiation Protection Program, is the primary federal agency charged with developing radiation protection standards for protecting the public and the environment from harmful and avoidable exposure to radiation.[9] The EPA also provides guidance and training to other federal and state agencies in preparing for emergencies at U.S. nuclear plants, transportation

---

7 See the OCRWM website for a detailed explanation of the agency's role in highly radioactive waste policy, including the Yucca Mountain project and waste acceptance and transportation [http://www.ocrwm.doe.gov].

8 See (NRC, 2005f) "Who We Are," U.S. Nuclear Regulatory Commission, [http://www.nrc.gov/who-we-are.html] and (NRC, 2005e) "What We Do," U.S. Nuclear Regulatory Commission [http://www.nrc.gov/what-we-do.html] for an overview of the Commission's role in radioactive waste policy.

9 See (EPA, 2000) "Radiation Protection at the EPA: The First 30 Years…Protecting People and the Environment," U.S. Environmental Protection Agency, Office of Radiation

accidents involving shipments of radioactive materials, and acts of nuclear terrorism. Moreover, the EPA is responsible for developing the environmental standards to evaluate the safety of a national geologic repository. (EPA, 2004: 1) Finally, the concept of operations in the Federal Radiological Emergency Response Plan (FRERP) designates the EPA as the lead federal agency for emergencies involving radiological material not licensed or owned by a federal agency or an agreement state. (FRERP, 1996: II-2)

The *Department of Transportation (DOT)* has the primary responsibility for establishing and enforcing standards for the shipment of highly radioactive waste. The DOT's Federal Motor Carrier Safety Administration (FMCSA) enforces radiological materials regulations such as the classification of the materials and proper packaging to ensure the safe and secure transportation by highway. (FMCSA, 2004) DOT's Pipeline and Hazardous Materials Safety Administration (PHMSA) oversees the safety of highway hazardous materials shipments including radiological waste. (PHMSA, 2004b) The PHMSA's Office of Hazardous Materials Safety oversees the Hazardous Materials Transportation Program, which is tasked to identify and manage risks presented by the transportation of hazardous material such as radioactive waste. (PHMSA, 2005b) DOT's Federal Railroad Administration (FRA) is the government agency responsible for monitoring highly radioactive waste shipments by rail. (DOE, 2002: 4) The FRA has developed a safety compliance plan to promote the safety of highly radioactive waste by rail.[10] (FRA, 1998: iv)

The *Federal Emergency Management Agency (FEMA)* also plays a role in nuclear waste policy.[11] For example, commercial nuclear power facilities in the United States are required to have both an on-site and off-site emergency response plan as a condition for obtaining and maintaining a license to operate the plant. While the NRC approves on-site emergency response plans, FEMA is responsible for evaluating off-site plans. The NRC must consider the FEMA findings when issuing or maintaining a license. (FEMA, 2004: 1) In addition, FEMA would play a crucial role in the response and recovery phase of any transportation accident involving nuclear waste that would result in the release of radioactivity.

The *Nuclear Waste Technical Review Board (NWTRB)* is an independent agency responsible for monitoring the U.S. radioactive waste management program. The NWTRB is tasked to provide independent scientific and technical oversight of the

---

and Indoor Air, EPA 402-B-00-001, August 2000, for a detailed description of the EPA's role in radioactive waste policy [http://www. epa.gov/radiation/docs/402-b-00-001.pdf].

10 See the DOT website for an overview of the department's role in radioactive waste policy [http://www.dot.gov]. The FMCSA website is available at: http://www.fmcsa.dot.gov. The PHMSA website is available at: http://www.phmsa.dot.gov. Information on the Hazardous Materials Transportation Program is available on the PHMSA's Office of Hazardous Materials Safety website at: http://hazmat.dot.gov/riskmgmt/risk.htm. See (FRA, 1998) "Safety Compliance Oversight Plan for Rail Transportation of High-Level Radioactive Waste & Spent Nuclear Fuel," Department of Transportation, Federal Railroad Administration, June 1998, [http://www.fra.dot.gov/downloads/safety/scopfnl.pdf] for a detailed description of the FRA's role in radioactive waste transportation.

11 See the FEMA website for an overview of the agency's role in radioactive waste policy [http://www.fema.gov].

U.S. program for the management and disposal of spent nuclear fuel and high-level radioactive waste (NWTRB, 2004: 1).

The Courts The role of the judicial branch in the U.S. political system is far greater today than initially anticipated by the Founders. The principle of *judicial review* (i.e., the ability of the courts to rule on the constitutionality of an issue) allows the courts to oversee both the legislative and executive branches and rule on the constitutionality of their actions. Moreover, the courts play an important role in policymaking since their rulings not only adjudicate conflict but ultimately establish policy. Today, the impact of the judicial branch is especially significant in contentious issues such as highly radioactive waste storage and disposal.

While the concept of federalism has given the U.S. a dual court system composed of both federal and state courts, it is the federal courts that have played the major role in adjudicating conflict over radioactive waste. This is due primarily to the nature of the radioactive waste issue. First, radioactive waste storage and disposal is national in scope, not just a state or local issue. Thus, the principle of national supremacy almost always guarantees that the federal courts will have the last word. Second, since spent nuclear fuel and high-level radioactive waste will have to be transported through multiple jurisdictions (both state and local) en route to a centralized interim storage facility or a permanent repository, the interstate commerce clause again ensures that the radioactive waste issue ultimately will be a federal question. Consequently, the federal courts will continue to be the courts of last resort for issues involving the storage and disposal of highly radioactive waste.

Since the delay in establishing either a centralized interim storage facility or a national repository for highly radioactive waste has served only to intensify conflict, the federal courts have played an increasingly active role in more recent years and have ruled in a variety of cases involving radioactive waste. For example, several states and utilities filed lawsuits against DOE to force the government to uphold its promise to accept nuclear waste by the original 1998 deadline. Moreover, various activist groups have filed lawsuits to prevent dry cask storage at particular sites, thereby making on-site storage for utilities a more difficult option. (NRC, 1996: 7) The courts have ruled on a number of cases involving interim radioactive waste storage in Utah and permanent disposal in Nevada. More recently, litigation has focused on the safety of the transportation of radioactive waste. Opponents of building a centralized radioactive waste storage or disposal facility now view the transportation issue as a primary means to block the construction of any centralized facility involving the storage or disposal of highly radioactive waste. In any case, the increased level of litigation has had the effect of delaying resolution of the problem of how best to cope with the mounting inventories of highly radioactive waste.

State/Tribal/Local Governments

Federal control of the nuclear waste debate has in effect created a means for some states to fair better than others in escaping the legacy of nuclear energy for well over 50-years. The doctrine of national supremacy in matters concerning radioactive waste is rooted in the Atomic Energy Act of 1954. The Act assigned responsibility for the disposal of highly radioactive waste to the federal government. (Flint, 2000: 2)

Although the federal government is preeminent in radioactive waste policy, the states, tribal governments, and local governments clearly can play an important role in the radioactive waste policy debate since they can be involved either as a host for a storage or disposal facility, or with the transportation of highly radioactive waste through their jurisdictions. Moreover, the federal government permits state, tribal, and local governments to pass legislation that specify requirements for transportation of radioactive materials within their jurisdictions as long as these laws are consistent with federal law, and do not impede interstate commerce. (DOE, 1999: 23)

State Governments A number of states have passed laws attempting to regulate or block the construction of radioactive waste storage/disposal facilities within their borders. For example, Utah passed several laws that eventually were overturned by a federal court which would, in effect, have made the operation of a centralized interim storage facility impossible (e.g., impose a $150 billion up-front bond for the facility, impose a 75 percent tax on anyone providing services to the project, bar the host county from providing municipal services such as fire and police, and restrict the transport of nuclear waste into the state). (Roche, 2002: 1) In all, almost 500 laws and regulations concerning radioactive waste have been passed by the states. Most of these laws principally have affected radioactive waste highway transport such as restrictions on routes, registration requirements, vehicle restrictions, and permits that affect the transportation of spent nuclear fuel. (Reed, 2000: 9)

State government officials basically have adopted one of three policy approaches to the nuclear waste issue:

- a majority of states have paid little or no attention to the problem
- some states have adopted a policy of active resistance
- some states have taken a proactive policy approach to regulating nuclear waste.

First, most states have adopted a wait and see attitude since they either have not been identified as a potential host for a storage or disposal facility, or are not yet directly affected by the transport of highly radioactive waste across their borders. As a rule, the nuclear waste question is not a salient issue with the public. As a result, state officials are usually reluctant to take on a challenging policy question when it is not an issue that the public expresses much concern over. Second, a few states have adopted a posture of active opposition to federal efforts to site an interim storage facility or a permanent repository. For example, Utah has attempted to block the Skull Valley interim storage initiative through legislation, litigation and a public relations campaign. Similarly, Nevada has strenuously opposed the Yucca Mountain national repository project and has had some success in delaying the project. In 2002, South Carolina unsuccessfully attempted to block high-level radioactive waste shipments from the Rocky Flats defense facility in Colorado to its Savannah River nuclear weapons complex. (Jordan, 2002: 1) Third, a small number of states have decided to adopt a proactive stance, especially in managing the nuclear waste that will come across their borders.[12] For example, Illinois has adopted a model radioactive waste transportation program. The state has had a long history with nuclear waste since a commercial nuclear waste reprocessing facility was constructed in Morris, Illinois in the 1970s. Although the facility never opened due to technical problems, it became the nation's first de facto interim storage facility when it was forced to accept spent nuclear fuel from commercial nuclear reactors that the reprocessing facility had contracted with to reprocess their waste. This necessitated transporting highly radioactive waste across the state to the reprocessing facility. New Mexico also has adopted a proactive approach, due to the construction of the Waste Isolation Pilot Plant (WIPP) near Carlsbad, which accepts transuranic waste from out-of-state defense industry facilities. State officials have opted to pass regulations that would ensure a more active role in monitoring and regulating the storage and transport of the radioactive waste within the state.

Governors also have been active in lobbying radioactive waste policy. For example, the Western Governors' Association (WGA) has been the most active of the five regional governors associations in the radioactive waste debate since several western states have been identified as potential hosts for a storage/disposal facility (e.g., WIPP – New Mexico, Yucca Mountain – Nevada, Skull Valley – Utah, and Owl Creek – Wyoming). The WGA has issued a number of policy resolutions opposed to establishing a private interim storage or disposal facility without state permission, as well as calling for commercial spent nuclear fuel to remain on-site until a permanent disposal facility is operational. (WGA, 2000: 2) The WGA also issued another policy resolution that requested the NRC to undertake a comprehensive reassessment of terrorism and sabotage risks for radioactive waste shipments and incorporate risk management and countermeasures in all DOE transportation plans. (WGA, 2001: 2-3) In 2002, the WGA reiterated that the commercial nuclear power industry and federal government should bear the costs associated with ensuring the safe transportation and effective response to accidents and emergencies that could occur. (WGA, 2002: 4) It should not be surprising that the governors of Nevada and Utah sponsored most of these resolutions since their states have been selected as possible hosts for a permanent highly radioactive waste repository (Yucca Mountain, Nevada) or a centralized interim spent fuel storage facility (Skull Valley, Utah).

While states have attempted to regulate radioactive waste, these laws must be consistent with federal laws, follow DOT guidelines, and not impede interstate commerce. Thus, it is clear that the principle of national supremacy and the interstate commerce clause will be used as the principal means to assert federal control over highly radioactive waste policy.

Tribal Governments Tribal governments have been active in the interim radioactive waste debate primarily due to opposition by the state governments to any locally sponsored centralized interim radioactive waste storage initiatives. Since governors

---

12 See (Reed, 2000) James B. Reed, "The State Role in Spent Fuel Transportation Safety: Year 2000 Update," National Conference of State Legislatures, January 2000, No. 14 [http://www.ncsl.org/programs/transportation/transer14.htm] for an excellent summary of state legislation and regulations issued on the transport of radioactive waste.

have blocked local initiatives, the stage was set for the federal government to begin negotiations with Native American tribes. In all, over 20 different Native American tribes have been involved at one stage or another in the voluntary siting process for centralized interim storage. The use of Native American tribal lands is an interesting choice with considerable political overtones. The main concern of state officials is that the use of tribal lands could circumvent state sovereignty and erode a state's ability to influence the decision making process. Indeed, the concept of tribal sovereignty (i.e., recognized as sovereign entities by the federal government) allows Native American tribes to circumvent state control on a number of issues such as gambling and taxes. (Gowda, 1998: 247) While there is considerable debate over the extent of tribal sovereignty with respect to radioactive waste storage, it does appear that the concept itself would be an attractive means to avoid opposition from state governments. Only after efforts to conclude an agreement with state and local governments failed, was there a more concerted effort to consult Native American tribes. (Gowda, 1998: 253)

In spite of the overtures to Native American tribes, no centralized interim radioactive waste storage facility was opened during the 1990s. However, the Skull Valley (Utah) project that is sponsored by a consortium of nuclear utilities had made considerable progress until adverse government rulings in 2006.

Local Governments Historically, local governments (counties, cities and townships) have been the least involved government entity in dealing with radioactive waste issues since the federal government has the overall primary responsibility for establishing standards and regulating nuclear waste, and state governments have the primary responsibility for monitoring radioactive waste shipments within state jurisdictional boundaries. The primary concern for local governments has been the emergency response capability to any potential accident involving the transport of radioactive materials. Local governments have attempted to influence policy by passing resolutions calling for the federal government to undertake certain actions. In 2002, the U.S. Conference of Mayors passed a resolution opposing the transportation of highly radioactive waste across state borders until the federal government can provide "adequate funding, training and equipment to protect public health and safety" for all cities along the proposed transportation route. (USCM, 2002: 1) In 2003, the National Association of Counties adopted a policy resolution that urged DOE to develop a transportation plan and accompanying environmental impact statement for highly radioactive waste shipments. (NACO, 2002: 25-26) While there clearly has been some concern displayed over the safety of the nuclear waste shipments – especially through communities with large population centers – it is difficult to measure the extent of the opposition since the issue is also being used by local officials as a vehicle to obtain additional funding from the federal government.

While local governments have attempted to play a larger role in the nuclear waste policy debate, they nevertheless have not had much of an impact on policy. The state-local government relationship is a unitary system arrangement where the local authorities only have the power granted by the state. Generally, state governments have been reluctant to delegate much authority in this area to local control. This is likely due to the potential for local governments to be more supportive (for economic reasons such as jobs, tax revenue, and storage or disposal fees) of establishing a disposal/storage facility within their jurisdiction. For example, several local governments received federal funding in the late 1980s or early 1990s to explore constructing a highly radioactive waste storage facility. However, the states generally blocked the initiative.[13] In 2002, county commissioners in Nye County, Nevada (where the proposed Yucca Mountain repository will be located) took steps to work more closely with DOE in order to have more influence over the process. (Tetreault, 2002b: 1) Their attempt to work with DOE ran counter to the state's desire to scuttle the project.

Interest Groups

Numerous interest groups have been active in the radioactive waste policy debate. They generally can be broken down into three major categories: (1) state and local government affiliated organizations; (2) environmental and citizen activist groups; and (3) the commercial nuclear power industry and supporters.[14] The state and local government affiliated organizations have focused on emergency response to a potential accident involving radiological materials. Environmental and citizen activist groups argue that the radioactive waste can continue to be stored on-site at the point of origin, thus alleviating the need to construct a storage or disposal facility as well as transport it to a centralized facility. In contrast, the commercial nuclear power industry has been a strong advocate of building either a centralized interim storage facility or a national repository for highly radioactive waste in order to avoid the premature shutdown of the reactors due to a lack of sufficient on-site radioactive waste storage capacity.

There are a number of state and local government affiliated organizations that have lobbied intensively on radioactive waste issues. For example, as discussed previously, the Western Governors' Association has issued several resolutions regarding spent fuel transportation and the National Association of Counties and the U.S. Mayors Conference adopted policy statements calling for more training and funding for local emergency response along the proposed transportation routes for the spent nuclear fuel. In addition, the Western Interstate Energy Board (an

---

13 The federal government offered funding incentives to explore the possibility of establishing a federally sponsored interim storage facility (known as a monitored retrievable storage – MRS – site). Two counties (Freemont County, WY and Grant County, ND) received preliminary federal grants that were eventually blocked by the state.

14 The Office of Civilian Radioactive Waste Management has cooperative agreements with a number of organizations interested in the transportation of spent nuclear fuel and highly radioactive waste: National Congress of American Indians (NCAI), National Association of Regulatory Utility Commissioners (NARUC), National Conference of State Legislatures (NCSL), The Council of State Governments/Eastern Regional Conference (CSG/ERC), The Council of State Governments-Midwestern Office (CSG-MW), Southern States Energy Board (SSEB), Western Interstate Energy Board (WIEB). There also were past agreements with the Commercial Vehicle Safety Alliance, Conference of Radiation Control Program Directors and the League of Women Voters Education Fund. (OCRWM, 2003: 1-3)

organization of western states and three Canadian provinces) has been critical of DOE and has reiterated the western governors' positions on the issues. (WIEB, 1996: 2; WIEB 1998: 1; and WIEB 2002:1) The National Association of Regulatory Utility Commissioners (NARUC), a non-profit organization whose members include government agencies engaged in regulating utilities and carriers in the fifty states, District of Columbia, Puerto Rico and the Virgin Islands also has issued a number of resolutions regarding highly radioactive waste.[15] In 2000, NARUC adopted a resolution supporting the development of both a centralized interim storage facility and a permanent repository so that commercial nuclear power facilities will not have to shut down prematurely due to a lack of sufficient on-site storage space. (NARUC, 2000: 1-2) In 2004 and 2005, NARUC passed resolutions supporting reform of the Nuclear Waste Fund so that monies paid into the fund by the commercial nuclear utilities are used for a permanent repository and not other activities. (NARUC, 2004: 1, NARUC, 2005: 1) The Conference of Radiation Control Program Directors (CRCPD), a non-profit professional organization dedicated to radiation protection, issues resolutions and position papers on radioactive waste management.[16] (CRCPD, 2005)

A number of groups opposed to either the centralized storage or disposal of highly radioactive waste or the transportation of the waste have been active in the debate. For example, the "NO Coalition" (a group of citizens and various political, business and former military leaders) has been especially active and has cited numerous environmental and safety concerns over the proposed Skull Valley centralized interim waste project.[17] (Bryson, 2002: 1) While there have been a number of groups opposed to the transportation of radioactive waste, the Transportation Safety Coalition has been especially active in attempting to block the Yucca Mountain project.[18] Some of the environmental and citizen activist opposition to either the construction of a centralized storage or disposal facility or the transport of highly radioactive waste is genuine due to safety and health concerns. However, other opposition groups hope eventually to shut down the commercial nuclear power industry entirely by preventing an away-from-the-reactor storage capability; thus, forcing the premature shutdown of reactors due to the limited availability of sufficient on-site storage space.

---

15 See the NARUC website for a description of the organization's mission and organization. [http://www.naruc.org/displaycommon.cfm?an=1].

16 The CRCPD issues the "Directory of State and Federal Agencies Involved with the Transportation of Radioactive Material," which is a useful resource for points of contact within state and federal agencies involved in radioactive waste management. (CRCPD, 2004)

17 For a detailed explanation of the Coalition's concerns, see (NO Coalition, 2000), "White Paper Regarding Opposition to the High-Level Nuclear Waste Storage Facility Proposed by Private Fuel Storage on the Skull Valley Band of Goshute Indian Reservation, Skull Valley, Utah," The Coalition Opposed to High-Level Nuclear Waste, 28 November 2000.

18 The Transportation Safety Coalition is composed of the American Public Health Association, the Environmental Working Group, the National Environmental Trust, Physicians for Social Responsibility, U.S. Public Information Research Groups, and the Nevada Agency for Nuclear Projects. (Hall, 2002: 1)

The commercial nuclear power industry and supporters understand that there will be an insufficient amount of on-site spent fuel storage capability in the near future and that a number of reactors will be forced to shut down if a suitable alternative for storing or disposing spent fuel is not found in the near-term. The Nuclear Energy Institute (NEI), which is the nuclear power industry's main lobbying organization, has been especially vocal and active in support of building a centralized storage or permanent disposal facility. Moreover, the NEI has stressed the positive safety record of transporting radioactive materials over the years. Private Fuel Storage (PFS), a consortium of eight commercial power utilities, also has been active in the debate since the announcement of their intention to build a centralized interim storage facility for spent fuel at Skull Valley, Utah. The Skull Valley facility would accept highly radioactive waste from existing nuclear power facilities so that the reactors will not be forced to shut down due to a lack of on-site storage space for the spent nuclear fuel. Storage and transportation cask vendors also have been active in the radioactive waste debate primarily citing the ability of the storage and transportation casks to withstand a substantial impact without releasing radiation. (NAC, 2003: 1-2)

Media

The media can play an important role in the nuclear waste policy debate by virtue of its gatekeeper role of determining what is news (i.e., deciding on what to publish). If the media decides to report on a controversial issue, then it will have a high likelihood of being put on the political agenda. For example, the extensive media coverage of the Three Mile Island and Chernobyl reactor accidents focused considerable attention on the safety of commercial nuclear power plants. As a result, public opposition to nuclear generated power rose significantly, not only preventing the construction of new plants, but also ultimately leading to the closing of several commercial nuclear power facilities.

In more recent years, there has been increased national coverage of the Yucca Mountain repository. Moreover, there has been a substantial amount of state and local reporting on the Skull Valley, Utah, and to a lesser extent Owl Creek (Wyoming), centralized interim storage initiatives. While the media has not yet focused much attention on the highly radioactive waste question, they could play an important *future* role in the debate. For example, unfair critical reporting by the media could serve to undermine public confidence in the safety of radioactive waste storage, disposal, or transportation. This, in turn, would likely increase citizen opposition to any storage or disposal plan. On the other hand, if the media focuses on the positive safety record and the low likelihood of an accident that could result in a release of radiological material, then citizen opposition will likely be minimized.

Summary

The problem of highly radioactive waste storage and disposal is a complex issue that involves security, political, economic, and societal issues. Thus, it is not an issue that is easily addressed. Moreover, since highly radioactive waste storage and disposal is not a salient topic with the public, there is little incentive for policymakers to address the problem since the costs are tangible and immediate, while the benefits are intangible and long-term. This, in turn, has delayed progress on developing a long-term solution to the problem. However, the inability of policymakers to develop a viable near-term solution to the problem of what to do with the ever-increasing amounts of radioactive waste being generated has both homeland security and energy security implications.

The events of 9/11 have focused attention on the vulnerability of nuclear power facilities – especially the highly radioactive waste stored on-site. At the same time, the lack of sufficient on-site storage space could force some commercial nuclear power facilities to shut down prematurely which could have a significant negative impact on energy security since it would increase U.S. reliance on foreign sources of energy. However, because of the conflict associated with the proposed centralized interim storage and permanent disposal initiatives, it has been easier for policymakers to continue the status quo policy of on-site storage at the reactor.

The inability of U.S. policymakers to develop an interim storage solution or a permanent disposal option has led to a de facto policy of storing highly radioactive waste on-site at the point of origin. Chapter 2 discusses the history of on-site storage, the types of storage being used today, and advantages and disadvantages of storing the spent nuclear fuel and high-level waste on-site.

Key Terms

storage
disposal
Nuclear Waste Policy Act of 1982 (NWPA)
Nuclear Waste Policy Amendments Act of 1987 (NWPAA)
homeland security
energy security
rolling stewardship
national supremacy
interstate commerce clause
Department of Energy (DOE)
Office of Civilian Radioactive Waste Management (OCRWM)
Nuclear Regulatory Commission (NRC)
Environmental Protection Agency (EPA)
Department of Transportation (DOT)
Federal Emergency Management Agency (FEMA)
Nuclear Waste Technical Review Board (NWTRB)

Useful Web Sources

National Association of Regulatory Utility Commissioners [http://www.naruc.org]
Nuclear Energy Institute [http://www.nei.org]

U.S. Department of Energy [http://www.doe.gov]

U.S. Department of Energy, Energy Information Administration [http://www.eia.doe.gov]

U.S. Department of Energy, Transportation External Working Group [http://www.nnsa.doe.gov/na-26/docs/ntp.pdf]

U.S. Department of Energy, Office of Civilian Radioactive Waste Management [http://www.ocrwm.doe.gov]

U.S. Department of Homeland Security, Federal Emergency Management Agency [http://www.fema.gov]

U.S. Department of Transportation, Federal Motor Carrier Safety Administration [http://www.fmcsa.dot.gov]

U.S. Department of Transportation, Federal Railroad Administration [http://www.fra.dot.gov]

U.S. Department of Transportation, Pipeline and Hazardous Materials Safety Administration [http://www.phmsa.dot.gov]

U.S. Environmental Protection Agency [http://www.epa.gov]

U.S. Nuclear Regulatory Commission [http://www.nrc.gov]

U.S. Nuclear Waste Technical Review Board [http://www.nwtrb.gov]

Western Governors' Association [http://www.westgov.org]

Western Interstate Energy Board [http://www.westgov.org/wieb]

References

(Abraham, 2002) Abraham, Spencer. Letter to President Bush recommending approval of Yucca Mountain for the development of a nuclear waste repository, Secretary of Energy, U.S. Department of Energy, 14 February 2002.

(Abrahms, 2003) Abrahms, Doug. "Reid cuts Yucca Mountain funding," *Reno-Gazette Journal*, 1/21/2003 [http://www.rgj.com/news/stories/html/2003/01/21/32573.php]

(Bryson, 2002) Bryson, Amy Joi. "Utah Group just says no to N-waste at Goshute," *Deseret News*, 27 April 2002.

(CRCPD, 2004) "Directory of State and Federal Agencies Involved with the Transportation of Radioactive Material," Conference of Radiation Control Program Directors, CRCPD Publication E-04-6, October 2004 [http://crcpd.org/Transportation/TransDir04.pdf]

(CRCPD, 2005) Conference of Radiation Control Program Directors [http://crcpd.org/Transportation_ related_docs.asp]

(DOE, 1999) "Transporting Radioactive Materials: Answers to Your Questions," National Transportation Program, Assistant Secretary for Environmental Management, U.S. Department of Energy, June 1999.

(DOE, 2002) "Transporting Spent Nuclear Fuel and High-Level Radioactive Waste to a National Repository: Answers to Frequently Asked Questions," U.S. Department of Energy, Office of Civilian Radioactive Waste Management, Yucca Mountain Project, December 2002 [http://www.ocrwm.doe.gov/wat/pdf/snf_transfaqs.pdf]

(DOE, 2004a) "About DOE: Mission," U.S. Department of Energy [http://www.doe.gov/engine/content.do?BT_CODE=ABOUTDOE]

(DOE, 2004b) "DOE Researchers Demonstrate Feasibility of Efficient Hydrogen Production from Nuclear Energy," Press Release, 30 November 2004 [http://www.doe.gov/news/1545.htm]

(DOE, 2004c) "Nuclear Power 2010," Office of Nuclear Energy, Science and Technology, U.S. Department of Energy, updated: 8/27/04 [http://nuclear.gov/NucPwr2010/NucPwr2010.html]

(EIA, 2004) "Annual Energy Outlook 2004," Energy Information Administration, U.S. Department of Energy, Table A8 (Electric Supply, Disposition, Prices and Emissions): 145 [http://www.eia.doe.gov/oiaf/archive/aeo04/pdf/0383(2004).pdf]

(EPA, 2000) "Radiation Protection at the EPA: The First 30 Years...Protecting People and the Environment," U.S. Environmental Protection Agency, Office of Radiation and Indoor Air, EPA 402-B-00-001, August 2000 [http://www.epa.gov/radiation/docs/402-b-00-001.pdf]

(EPA, 2004) "About EPA's Radiation Protection Program," Radiation Information, U.S. Environmental Protection Agency [http://www.epa.gov/radiation/about/index.html]

(Falcone and Orosco, 1998) Falcone, Santa and Kenneth Orosco. "Coming Through a City Near You: The Transport of Hazardous Wastes," *Policy Studies Journal*, vol. 26, Issue 4, Winter 1998: 760-73.

(FEMA, 2004) "Backgrounder: Nuclear Power Plant Emergency," Hazards, Federal Emergency Management Agency [http://www.fema.gov/hazards/nuclear/radiolo.shtm]

(Flint, 2000) Flint, Lawrence. "Shaping Nuclear Waste Policy at the Juncture of Federal and State Law," *Boston College Environmental Affairs Law Review*, vol. 28, No. 1, 2000 [http://www.bc.edu/schools/law/lawreviews/meta-elements/journals/bcealr/28_1/05_FMS.htm]

(FMCSA, 2005) "About Us," Department of Transportation, Federal Motor Carrier Safety Administration [http://www.fmcsa.dot.gov/aboutus/aboutus.htm]

(FRA, 1998) "Safety Compliance Oversight Plan for Rain Transportation of High-Level Radioactive Waste & Spent Nuclear Fuel," U.S. Department of Transportation, Federal Railroad Administration, June 1998, [http://www.fra.dot.gov/downloads/safety/scopfnl.pdf]

(FRA, 2005) "Hazardous Materials Division," Federal Railroad Administration, U.S. Department of Transportation [http://www.fra.dot.gov/us/content/337]

(FRERP, 1996) "Federal Radiological Emergency Response Plan," May 1, 1996. Summary available in the *Federal Register*, vol. 61, No. 90, Wednesday, 8 May 1996, Notice: 20 944.

(Gowda, 1998) Gowda, M.V. Rajeev and Doug Easterling. "Nuclear Waste and Native America: The MRS Siting Exercise," *Risk Health, Safety and Environment*, Summer 1998: 229-58.

(GPO, 2007) "Budget of the United States Government: Fiscal Year 2008," U.S. Government Printing Office, 2007 [http://www.gpoaccess.gov/usbudget/fy08/pdf/budget/energy.pdf]

(Hall, 2002) Hall, Jim. "Testimony of Jim Hall on Behalf of the Transportation Safety Coalition," U.S. Senate, Committee on Energy and Natural Resources, 23 May 2002 [http://www.yuccamountain.org/pdf/hall052302.pdf]

(Jordan, 2002) "S.C. to block plutonium trucks," *Arkansas Democrat Gazette*, 15 June 2002, p. 2A, col. 5.

(Katz, 2001) Katz, Jonathan L. "A Web of Interests: Stalemate on the Disposal of Spent Nuclear Fuel," *Policy Studies Journal*, vol.29, Issue 3, 2001: 456-77.

(McCutcheon, 2002) McCutcheon, Chuck. *Nuclear Reactions: The Politics of Opening a Radioactive Waste Disposal Site*, University of New Mexico Press, 2002.

(Meserve, 2002) Meserve, Richard A. Letter to Congressman Edward J. Markey, 4 March 2002 [http://www.house.gov/markey/ISS_nuclear_ltr020325a.pdf]

(Munton, 1996) Munton, Don. "Introduction: The NIMBY Phenomenon and Approaches to Facility Siting," in *Hazardous Waste Siting and Democratic Choice: The NIMBY Phenomenon and Approaches to Facility Siting*, Georgetown University Press, 1996.

(NAC, 2003) "Energy Solutions, Information and Technology," NAC International, 2003 [http://www.nacintl.com]

(NACO, 2002) "Environment, Energy & Land Use," *The American County Platform and Resolutions*, Resolution 3A, Adopted July 16, 2002 [http://www.naco. org/Content/ContentGroups/Legislative_Affairs/Advocacy1/EELU/EELU1/ 3EELU_02-03.pdf]

(NARUC, 2000) "Resolution Regarding Guiding Principles for Disposal of High-Level Nuclear Waste," National Association of Regulatory Utility Commissioners, 2000 Resolutions and Policy Positions (Electricity) [http://www.naruc.org/ Resolutions/2000/annual/elec/disposal_nuclear_waste.shtml]

(NARUC, 2004) "Resolution Supporting Reform of the Nuclear Waste Fund," National Association of Regulatory Utility Commissioners, 2004 Resolutions and Policy Positions (Electricity) [http://www.naruc.org/displaycommon.cfm?an= 1&subarticlenbr=295]

(NARUC, 2005) "Resolution Supporting Reform of the Nuclear Waste Fund," National Association of Regulatory Utility Commissioners, 2005 Resolutions and Policy Positions: Electricity [http://www.naruc.org/displaycommon.cfm?an= 1&subarticlenbr=394]

(NAS, 1957) "The Disposal of Radioactive Waste On Land," Report of the Committee on Waste Disposal of the Division of Earth Sciences, National Academy of Sciences, National Research Council, Publication 519, September 1957 [http:// www.nap.edu/books/NI000379/html]

(NAS, 2001) "Disposition of High-Level Waste and Spent Nuclear Fuel: The Continuing Societal and Technical Challenges," Committee on Disposition of High-Level Radioactive Waste Through Geological Isolation, Board on Radioactive Waste Management, Division on Earth and Life Studies, National Academy of Sciences, National Research Council, National Academy Press, Washington D.C., 2001 [http://www.nap.edu/openbook/0309073170/html/R1.html]

(NEI, 2004) "Powering Tomorrow…With Clean, Safe Energy," Vision 2020, Nuclear Energy Institute [http://www.nei.org/documents/Vision2020_Backgrounder_Powering_Tomorrow.pdf]

(NO Coalition, 2000), "White Paper Regarding Opposition to the High-Level Nuclear Waste Storage Facility Proposed by Private Fuel Storage on the Skull Valley Band of Goshute Indian Reservation, Skull Valley, Utah," The Coalition Opposed to High-Level Nuclear Waste, 28 November 2000.

(NRC, 1996) "Strategic Assessment Issue: 6. High-Level Waste and Spent Fuel," U.S. Nuclear Regulatory Commission, 16 September 1996 [http://www.nrc.gov/NRC/STRATEGY/ISSUES/dsi06isp.htm]

(NRC, 2005a "Emergency Preparedness and Response," U.S. Nuclear Regulatory Commission [http://www.nrc.gov/what-we-do/emerg-preparedness.html]

(NRC, 2005b) "Nuclear Materials," U.S. Nuclear Regulatory Commission [http://www.nrc.gov/materials.html]

(NRC, 2005c) "Nuclear Reactors," U.S. Nuclear Regulatory Commission [http://www.nrc.gov/reactors.html]

(NRC, 2005d) "Radioactive Waste," U.S. Nuclear Regulatory Commission [http://www.nrc.gov/waste.html]

(NRC, 2005e) "What We Do," U.S. Nuclear Regulatory Commission [http://www.nrc.gov/what-we-do.html]

(NRC, 2005f) "Who We Are," U.S. Nuclear Regulatory Commission [http://www.nrc.gov/who-we-are.html]

(NTP, 2005) "National Transportation Program Products & Services," National Transportation Program, U.S. Department of Energy [http://www.ntp.doe.gov/mission.html]

(Nuke-Energy, 2002) "OCRWM Chief Says Something: 2010 Milestone is 'Tight'," *Nuke-Energy*, July 2002 [http://www.nuke-energy.com/data/stories/July02/Chu.htm]

(NWTRB, 2004) "What is the U.S. Nuclear Waste Technical Review Board?" U.S. Nuclear Waste Technical Review Board, NWTRB Viewpoint, November 2004 [http://www.nwtrb.gov/mission/nwtrb.pdf]

(OCRWM, 2003) "Cooperative Agreements," U.S. Department of Energy, Office of Civilian Radioactive Waste Management [http://www.ocrwm.doe.gov/wat/cooperative_agreements.shtml]

(OCRWM, 2004a) "OCRWM Program Briefing," Office of Civilian Radioactive Waste Management, U.S. Department of Energy [http://www.ocrwm.doe.gov/pm/programbrief/briefing.htm]

(OCRWM, 2004b) "Overview," Office of Civilian Radioactive Waste Management, U.S. Department of Energy [http://www.ocrwm.doe.gov/overview.shtml]

(OCRWM, 2004c) "The U.S. Department of Energy's Role in the Nuclear Regulatory Commission's Licensing Process for a Repository," Office of Civilian Radioactive Waste Management, [http://www.ocrwm.doe.gov/factsheets/doeymp0113.shtml]

(OCRWM, 2005a) "Appropriations by Fiscal Year [Yucca Mountain]," Office of Civilian Radioactive Waste Management Program [http://www.ocrwm.doe.gov/pm/budget/budget.shtml]

(OCRWM, 2005b) "FY 2006 Budget Request Summary [Yucca Mountain]," Office of Civilian Radioactive Waste Management Program [http://www.ocrwm.doe. gov/pm/budget/budgetrollout_06/2006cbr7.shtml]

(OMB, 2004) "Budget of the United States – FY 2005," Office of Management and Budget, [http://www.whitehouse.gov/omb/budget/fy2005/pdf/budget/energy.pdf]

(OMB, 2005) "Budget of the United States – FY 2006," Office of Management and Budget, [http://www.whitehouse.gov/omb/budget/fy2006/pdf/budget/energy.pdf]

(PHMSA, 2005a) "About PHMSA," Pipeline and Hazardous Materials Safety Administration, U.S. Department of Transportation [http://www.phmsa.dot.gov/ about/index.html]

(PHMSA, 2005b) "Hazardous Materials Safety," Office of Hazardous Materials Safety, Pipeline and Hazardous Materials Safety Administration, U.S. Department of Transportation [http://www.phmsa.dot.gov/riskmgmt/risk.htm]

(PL109-58, 2005) "Energy Policy Act of 2005," Public Law 109-58, 8 August 2005 [http://frwebgate.access.gpo.gov/cgi-bin/getdoc.cgi?dbname=109_cong_public _laws&docid=f:publ058.109.pdf]

(Reed, 2000) Reed, James B. "The State Role in Spent Fuel Transportation Safety: Year 2000 Update," NCSL Transportation Series, January 2000, No. 14 [http:// www.ncsl.org/programs/transportation/transer14.htm]

(Roche, 2002) Roche, Lisa Riley. "Judge won't rule on wisdom of N-waste facility," Desert News, April, 2002 [http://www.energy-net.org/IS/EN/NUKE/WST/FED/ NEWS/NP-1212.402]

(TEC/WG, 2002) "Charter," The Transportation External Coordination Working Group, National Transportation Program, U.S. Department of Energy, Revised 2002 [http://www.tecworkinggroup.org/tecchart.pdf]

(TEC/WG, 2007) "TEC Members," Transportation External Working Group, U.S. Department of Energy [http://www.tecworkinggroup.org/members.html]

(Tetreault, 2002) Tetreault, Steve. "Nye County seeks role in nuclear waste project," Las Vegas Review-Journal, February 9, 2002. Available at [http://www. reviewjournal. com].

(Tetreault, 2003) Tetreault, Steve. "Congress settles on budget for Yucca Mountain," Las Vegas Review-Journal, 6 November 2003. [http://www.reviewjournal.com/ lvrj_home/2003/Nov-06-Thu-2003/news/22528355.html]

(USCM, 2002) "Resolution: Transportation of High-Level Nuclear Waste," U.S. Conference of Mayors, Adopted at the 70th Annual Conference of Mayors, Madison, WI, 14-18 June 2002.

(WGA, 2000) "Policy Resolution 00-031 – Private Storage of Commercial Spent Nuclear Fuel," Western Governors' Association, Policy Resolution 00-031, 13 June 2000.

(WGA, 2001) "Assessing the Risks of Terrorism and Sabotage Against High-Level Nuclear Waste Shipments to a Geologic Repository or Interim Storage Facility," Western Governors' Association, Policy Resolution 01-03, August 14, 2001 [http://www.westgov.org/wga/policy/01/01_03.pdf] Originally adopted as Policy Resolution 98-008.

(WGA, 2002) "Transportation of Spent Nuclear Fuel and High-Level Radioactive Waste," Western Governors' Association, Policy Resolution 02-05, June 25, 2002 [http://www.westgov.org/wga/policy/02/nuketrans_05.pdf]

(White House, 2001) "National Energy Policy," Report of the National Energy Development Group, May 2001 [http://www.whitehouse.gov/energy/Forward.pdf]

(WIEB, 1996) "HLW Committee Comments on DOE's Notice of Proposed Policy and Procedures for Safe Transportation and Emergency Response Training for Spent Nuclear Fuel and High-Level Radioactive Waste (Notice)," Western Interstate Energy Board, 12 September 1996 [http://www.westgov.org/wieb/reports/180c1998.htm]

(WIEB, 1998) "HLW Committee Comments on DOE's April 30, 1998 Notice of Revised Policy and Procedures for Safe Transportation and Emergency Response Training; Technical Assistance and Funding for Spent Nuclear Fuel and High-Level Radioactive Waste (Notice)," Western Interstate Energy Board, 31 July 1998 [http://www.westgov.org/wieb/reports/180c1998.htm]

(WIEB, 2002) "Reports and Comments," Western Interstate Energy Board, website [http://www.westgov.org/wieb/reports.html]

Chapter 2

On-Site Storage

Since the first controlled nuclear chain reaction occurred in 1942, considerable radioactive waste has been generated around the world. By 2005, over 250,000 metric tons of highly radioactive waste had been generated worldwide. By 2010, this amount is expected to almost double with an additional 210,000 metric tons being produced. (EIA, 2001: 1) Since the United States generates more highly radioactive waste than any other country, the problem of how best to cope with the mounting inventories of the lethal waste is more acute. Today, spent nuclear fuel from commercial nuclear power facilities and high-level waste from defense activities are now stored on-site at numerous locations around the country.

This chapter provides an overview of the types of radioactive waste that is generated; discusses the history of on-site highly radioactive waste storage; examines the spent fuel pool and dry cask storage methods; and provides an assessment of the advantages and disadvantages associated with storing the spent nuclear fuel and high-level waste on-site at the point of origin.

Radioactive Waste and its Impact

Highly radioactive waste is a by-product of the fission process of nuclear material. While nuclear fuel assemblies vary according to the type of reactor used (see Figure 2.1 – Nuclear Fuel Assembly for a depiction of a typical fuel assembly), a typical reactor uses solid ceramic pellets of enriched uranium that are sealed in metal tubes and then bundled together to form a nuclear fuel assembly. After three to five years in a reactor, the fuel assembly loses its ability to efficiently produce energy (i.e., it is "spent") and must be replaced. (OCWRM, 2004a: 6) The "spent" fuel assembly, which is highly radioactive and still generates significant heat, is then transferred to a spent fuel pool to cool slowly. After sufficient cooling, the spent nuclear fuel can be transferred to a dry cask container for continued on-site storage. Eventually, after sufficient cooling, the highly radioactive waste can be transferred to a centralized interim storage facility or permanent disposal repository.

There are two distinct types of highly radioactive waste that eventually will be stored in a centralized interim storage facility or permanent national repository: spent nuclear fuel, and high-level waste. *Spent nuclear fuel* is the by-product of the fission process from reactors in commercial nuclear power facilities, nuclear submarines and ships, and university and government research facilities. The spent nuclear fuel is a highly radioactive waste that contains plutonium as well as unconsumed uranium. (NAS, 2001: 8) Spent nuclear fuel is stored on-site near the reactor. *High-level waste* refers to the highly radioactive waste generated by DOE as a result of nuclear weapons production (e.g., chemical reprocessing of spent nuclear fuel to recover plutonium), surplus plutonium from dismantled nuclear weapons, and defense research and development programs. High-level liquid waste is stored in barrels on-site at the DOE complex. According to DOE, all high-level waste will be vitrified in solid form (i.e., mixed with glass or ceramic material) prior to being shipped to a repository. (OCRWM, 2004a: 7) The distinction between spent nuclear fuel and high-level waste is significant for homeland security concerns (e.g., proliferation) since the uranium and plutonium in spent nuclear fuel can be used either for future energy generation, or for nuclear weapons production. On the other hand, most high-level waste does not contain much uranium or plutonium, and thus while still a potential security vulnerability (e.g., use in a "dirty bomb"), is not much of a nuclear weapons proliferation concern. (NAS, 2001: 9) The term *highly radioactive waste* is used to describe both spent nuclear fuel generated from commercial nuclear power plants and high-level waste produced at DOE facilities. There also is *transuranic waste* (i.e., heavier than uranium) that consists mainly of contaminated protective clothing, rags, tools and equipment, chemical residues and scrap materials resulting from defense activities such as nuclear weapons production. Transuranic waste is disposed of at the Waste Isolation Pilot Plant (WIPP) in Carlsbad, New Mexico. Some of these radioactive wastes have a half-life (i.e., the amount of time for the radioactive isotopes to decay by half) of thousands of years; and thus, will be a threat to health and the environment for quite some time.

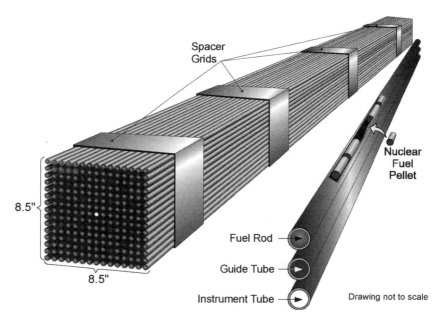

Figure 2.1 Nuclear Fuel Assembly

Source: "Spent Nuclear Fuel", OCRWM Program Briefing, Office of Civilian Radioactive Waste Management, U.S. Department of Energy [http://www.ocrwm.doe.gov/pm/programbrief/briefing.htm

History of On-Site Storage

Shortly after it was demonstrated that electricity could be produced from a nuclear reactor, Congress passed the 1954 Atomic Energy Act, which provided guidance and support for the development of commercial nuclear generated power. In 1957, the first commercial nuclear power plant began producing electricity. The *Price-Anderson Act* was enacted into law the same year to provide liability protection for reactor operators in the event of an accident that results in a catastrophic release of radiation.[1] Although the Act initially was passed to encourage nuclear power production, it also provides coverage for the highly radioactive waste stored on-site. In subsequent years additional commercial nuclear power plants were constructed; and today, nuclear generated power accounts for about twenty percent of the overall U.S. electricity production. At the same time commercial nuclear power generation expanded, DOE generated high-level waste as a result of its defense activities. In the early years, the amount of spent nuclear fuel and high-level waste generated was small and not considered to be much of a problem. However, as the years passed it became evident that a national policy was needed to cope with the growing stockpiles of highly radioactive waste being generated.

In 1982, the Nuclear Waste Policy Act (NWPA) began the process of attempting to find a solution to the problem of highly radioactive waste disposal for spent nuclear fuel from commercial nuclear power plants and high-level waste from defense complexes. Five years later, the 1987 Nuclear Waste Policy Act Amendments (NWPAA) singled out Yucca Mountain, Nevada, as the only site to be studied for a national repository for spent nuclear fuel and high-level waste. In 1998, the federal government missed the deadline to begin transferring spent nuclear fuel and high-level defense waste to a permanent repository. In 1999, legislation was introduced into the 106th Congress that would allow DOE to "take title" to the highly radioactive waste being stored on-site to relieve the liability and costs associated with on-site storage. The legislation eventually was withdrawn due to opposition from states that were concerned that this would remove the pressure on DOE to move aggressively on constructing a permanent repository. (Flint, 2000: 178-79) In 2000, DOE concluded an agreement with a commercial nuclear facility to nominally "take title" to the waste being stored on-site. This agreement was to serve as a framework for subsequent agreements that would allow utilities to reduce future payments into the Nuclear Waste Fund by the amount of costs incurred with continuing on-site storage. (DOE, 2000: 1) However, a legal challenge by the nuclear power industry forced cancellation of the agreement in 2002. Today, highly radioactive waste is still being stored on-site at the point of origin. While the federal government was supposed to take title to the commercial spent nuclear fuel in January 1998, the conflict generated by the site selection process as well as the geologic and environmental concerns

---

1 See (Callan, 1998) Callan, Joseph L. "NRC's Report to Congress on the Price-Anderson Act," U.S. Nuclear Regulatory Commission, SECY-98-160, July 2, 1998 [http://www.nrc.gov/reading-rm/doc-collections/commission/secys/1998/secy1998-160/1998-160scy.html] for a discussion of the Price-Anderson Act.

Calculated Risks

associated with the Yucca Mountain site has delayed construction of the federal repository (see Table 2.1 – History of On-Site Radioactive Waste Storage).

Table 2.1 History of On-Site Radioactive Waste Storage

1942–First controlled nuclear chain reaction occurs. Defense activities begin to produce high-level waste.

1954–Congress passes the Atomic Energy Act allowing commercial utilities to produce nuclear generated power.

1957–First commercial nuclear power plant becomes operational and begins to produce spent nuclear fuel.

1960s–Spent nuclear fuel accumulates on-site in spent fuel cooling pools.

1970s–Attempts to a site a centralized interim storage facility falter.

1982–Nuclear Waste Policy Act (NWPA) mandates the process of establishing a deep, underground permanent repository for spent nuclear fuel and high-level waste.

1986–First dry cask container for on-site storage certified by NRC for use.

1987–NWPA of 1982 amended to single out Yucca Mountain as the only site to be studied for permanent highly radioactive waste repository. Referred to as NWPAA.

1997/1998–Congressional attempts to allow DOE to construct a centralized interim storage facility at Nevada Test Site blocked by Clinton Administration.

1998–DOE fails to take title to the highly radioactive waste stored on-site as mandated by the 1982 NWPA.

2000–Congressional attempts to have DOE take title to on-site nuclear waste until permanent repository ready blocked by Clinton Administration.

2006–Spent nuclear fuel and high-level waste continues to be stored on-site since no centralized storage/disposal facility ready to accept highly radioactive waste.

It is not clear when Yucca Mountain could be ready to accept spent nuclear fuel and high-level waste. Although 2010 was the most often date cited for completion of the repository, it is doubtful that it will be open before 2015-2020 at the earliest. At the same time, attempts to find a centralized interim storage solution also have not been successful. Thus, the inability of U.S. policymakers to devise either a long-term or an interim solution to the problem of highly radioactive waste storage or disposal has led to a de facto policy of storing the nuclear waste on-site at the point of origin.

Today, over half of the U.S. population lives within a few miles of one of these sites. According to the Office of Civilian Radioactive Waste Management (OCRWM), commercial nuclear power reactors have generated over 50,000 metric tons of spent nuclear fuel now stored temporarily on-site near the reactors. (OCRWM, 2004a: 4) In addition to the commercial spent nuclear fuel, highly radioactive waste from defense facilities, nuclear naval ships, and surplus plutonium from nuclear weapons production, have piled up at numerous DOE sites across the country. Overall, at one time or another, DOE has had responsibility for cleaning up 146 sites in 34 states and Puerto Rico. (EM, 2004)

The delay in opening a national repository has had a significant impact on future on-site inventories of highly radioactive waste. The Yucca Mountain repository is scheduled to accept about 3,000 metric tons of spent nuclear fuel and high-level waste annually. At the same time, about 2,000 metric tons of highly radioactive waste is produced each year across the country. While spent nuclear fuel production is expected to decline as existing commercial nuclear power plants reach the end of their useful life- cycle and shut down, the Bush Administration's plan to build additional nuclear power plants as part of its overall energy policy will create additional spent nuclear fuel. Consequently, if the Yucca Mountain repository does not begin operations until 2015-2020 (the likely opening date of the repository) it is estimated that 80,000 metric tons (10,000 metric tons beyond the planned disposal capacity of the Yucca Mountain facility) of spent nuclear fuel will require storage. (NWTRB, 1996a: 11) Assuming that the additional 10,000 metric ton capacity was available, it would take about 80 years to eliminate all the on-site inventories of highly radioactive waste. Since any delay in opening up the repository increases storage needs by about 2,000 metric tons per year, any further delay in opening Yucca Mountain will have a significant impact on future highly radioactive waste inventories with concomitant homeland security and energy security implications. (NWTRB, 1996a: 12) By 2035, OCRWM projects that the U.S. highly radioactive waste totals will double to over 100,000 metric tons (see Table 2.2 – Projected U.S. Highly Radioactive Waste Quantities Through 2035). (OCRWM, 2004a: 4) In spite of the fact that highly radioactive waste continues to pile up at sites across the country, there has been a surprising lack of consensus on how best to address the problem.

Table 2.2 Projected U.S. Highly Radioactive Waste Quantities Through 2035

Commercial Spent Fuel	105,000 metric tons
DOE Spent Fuel	2,500 metric tons
Naval reactor fuel	65 metric tons
Research fuel loaned to other countries	16 metric tons
Surplus plutonium	50 metric tons
High-Level Radioactive Waste as Glass	22,280 canisters

Source: "Nuclear Waste–A Long-Term National Challenge," OCRWM Program Briefing, Office of Civilian Radioactive Waste Management, U.S. Department of Energy [http://www.ocrwm.doe.gov/pm/programbrief/briefing.htm]

Types of On-Site Storage

There are two primary types of *on-site storage* for spent nuclear fuel: spent fuel pools and dry cask containers. Spent nuclear fuel removed from the reactor is highly radioactive and emits intense heat. Therefore, the spent fuel rods must be transferred immediately to a spent fuel pool to cool slowly and provide the necessary shielding to protect against radiation exposure. After sufficient cooling, the spent fuel rods can be transferred to dry cask storage. High-level defense waste is stored on-site at the various weapons production facilities.

Spent Fuel Pools

Spent fuel pools are concrete and steel lined pools located near the reactor and designed to provide radioactive shielding and cooling for the spent fuel rods (see Figure 2.2 – Spent Fuel Pool for a photo of a typical cooling pool). A typical spent fuel pool is around 20 feet deep and housed within a non-reinforced building. Because storage space in the cooling pools is limited and approaching maximum capacity in many pools, the utilities received permission to "re-rack" (i.e., move closer together) the spent fuel rods in the cooling pool. While *re-racking* the spent fuel rods has generated a significant increase in spent nuclear fuel storage capacity, as time has passed the cooling pools at some sites have reached their storage capacity.[2]

Figure 2.2 Spent Fuel Pool
Source: "Spent Fuel Pools." U.S. Nuclear Regulatory Commission [http://www.nrc.gov/waste/spent-fuel-storage/pools.html]

---

2   The spent fuel cooling pools cannot be totally filled since each reactor must maintain enough storage capacity in the cooling pool to be able to off-load all the active reactor rods in the event that an emergency shutdown becomes necessary.

Another option to expand in-pool storage is *rod consolidation*, where the fuel rods are removed from the spent fuel assembly and rearranged more compactly inside a metal canister. (OCRWM, 1995: 22) Although re-racking and rod consolidation have increased the storage capacity in the spent fuel pools, it is estimated that 78 reactors will have exhausted their spent fuel pool radioactive waste storage capacity by 2010. (NEI, 2004: 4) Thus, the utilities have been forced to search for an alternative method for on-site storage (see Figure 2.3 – Nuclear Spent Fuel Pool Capacity for a graphic depiction of the progressive loss of spent pool storage).

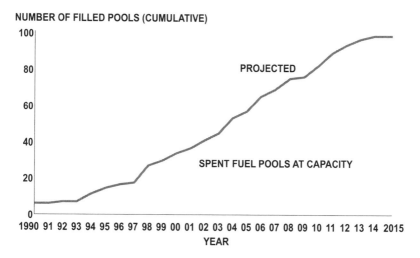

Figure 2.3 Nuclear Spent Fuel Pool Capacity
Source: "Nuclear Fuel Pool Capacity," U.S. Nuclear Regulatory Commission [http://www. nrc.gov/waste/spent-fuel-storage/nuc-fuel-pool.html]

Dry Cask Storage

Due to a lack of sufficient storage space in the spent fuel pools, a dry cask container storage system was developed to provide additional on-site storage capability. *Dry casks* are steel and concrete containers that use a passive ventilation method to enable the spent fuel rods to continue to cool slowly. After the casks are loaded with the spent fuel rods, they are filled with inert gas and sealed. The casks either can be stored vertically on a concrete pad, or horizontally in a reinforced concrete and steel storage module that resembles a vault or bunker (see Figure 2.4 – Dry Cask Storage of Spent Fuel for depictions of the dry cask storage methods).

The first dry cask container for on-site storage was certified by the NRC in 1986. Since the containers are certified for a period of 20 years, some of the earliest certified casks are nearing the end of the period of certification. Their license either must be extended, or the spent fuel will have to be transferred to another container. Either a site-specific license or a general license can be granted for on-site dry cask storage (see Figure 2.5 – Locations of Independent Spent Fuel Storage Installations). A *site-specific license* allows the nuclear utility to store waste off-site (away from the reactor) as well as accept spent fuel from other plants. For a site-specific license, the applicant must demonstrate that all the technical, administrative, and environmental licensing requirements authorizing the construction and operation of an independent spent fuel storage installation (away from the reactor site) meets all NRC requirements. (NEI, 1998: 62) A *general license* also can be granted to nuclear plant licensees to store spent fuel in NRC approved dry casks. A general license does not require the lengthy process of initiating a site specific license application requiring a safety analysis report, emergency plan, decommissioning plan, security plan, and environmental report. (NEI, 1998: 62) While a general license is less restrictive, it only allows the plant to store spent fuel that it is authorized to possess under its power reactor license (Raddatz and Waters, 1996: 2-3).

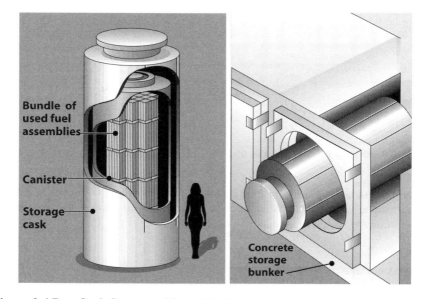

Figure 2.4 Dry Cask Storage of Spent Fuel
Source: "Dry Cask Storage of Spent Fuel," Backgrounder, U.S. Nuclear Regulatory Commission, p. 3 [http://www.nrc.gov/reading-rm/doc-collections/fact-sheets/dry-cask-storage.html

Today, there are a number of different types of NRC approved dry cask containers (see Table 2.3 – NRC Approved Dry Spent Fuel Storage Designs Currently in Use for a list of approved containers for highly radioactive waste and their storage location). However, there are some concerns with the dry casks storage containers. For example, safety incidents have occurred (e.g., defective welds, cracked seals, and explosions) with some of the casks. In addition, there has been concern expressed over the potential for sabotage or theft of the dry casks that are stored above ground

Figure 2.5 Locations of Independent Spent Fuel Storage Installations
Source: "Locations of Independent Spent Fuel Storage Installations," U.S. Nuclear Regulatory Commission [http://www.nrc.gov/waste/spent-fuel-storage/locations.html]

on a concrete pad. (NEI, 1998: 74-7; Macfarlane, 2001a: 218) For these reasons, the on-site dry cask storage option has become increasingly controversial. As the U.S. Nuclear Regulatory Commission (NRC) has pointed out:

> ...negative public response to dry storage facilities at some reactor sites has resulted in close scrutiny of dry cask storage activities, which, at times, has slowed the licensees'

Calculated Risks

progress toward implementing licensed dry storage operations. Various activist groups have filed lawsuits to prevent dry storage at particular sites and, at times, have influenced State and local governments, thereby making dry storage for utilities a more difficult option. [NRC, 1996: 7]

Thus, increased opposition to dry cask storage can be expected in the future.

Table 2.3 NRC Approved Dry Spent Fuel Storage Designs Currently in Use

Model (Storage Design)	Vendor	Date Approved (+ = general license)	Facilities Where Used (* = specific license)
CASTOR V/21 & X133 (Vertical Metal Cask)	General Nuclear Systems, Inc.	7/2/1986 8/17/1990+	Surry* (VA)
Fuel Solutions (Vertical Metal/Concrete Cask)	BFNL Fuel Solutions	2/15/2001+	Big Rock Point (MI)
HI-STAR 100 (Vertical Metal Cask)	Holtec International	10/4/1999+	Hatch (GA) Dresden (IL)
HI-STORM 100 (Vertical Metal/Concrete Cask)	Holtec International	3/31/1999 5/31/2000+	Trojan* (OR) Hatch (GA) Dresden (IL) Columbia (WA) FitzPatrick (NY) Arkansas Nuclear One (AR)
NAC-I28 (Vertical Metal Cask)	NAC International	2/1/1990	Surry* (VA)
NAC-UMS (Vertical Metal /Concrete Cask)	NAC International	11/20/2000+	Maine Yankee (ME) Palo Verde (AZ)
NAC-MPC (Vertical Metal /Concrete Cask)	NAC International	4/10/2000+	Yankee Rowe (MA) Haddam Neck (CT)
Advanced NUHOMS-24 (Horizontal Concrete Module)	Transnuclear, Inc.	02/05/2003+	San Onofre (CA)
NUHOMS (Horizontal Concrete Module)	Transnuclear, Inc.	8/13/1986 1/29/1990 11/25/1992 6/30/2000 1/18/1995+	H.B. Robinson* (SC) Oconee* (SC) Calvert Cliffs* (MD) Rancho Seco* (CA) Davis-Besse (OH) Susquehanna (PA) Duane Arnold (IA) Oyster Creek (NJ) Palisades (MI) Point Beach (WI)
TN-32 (Vertical Metal Cask)	Transnuclear, Inc.	7/2/1986 6/30/1998 4/19/2000+	Surry* (VA) North Anna* (VA) McGuire (NC) Peach Bottom (PA)
TN-40 (Vertical Metal Cask)	Transnuclear, Inc.	10/19/1993	Prairie Island* (MN)
TN-68 (Vertical Metal Cask)	Transnuclear, Inc.	5/28/2000+	McGuire (NC) Peach Bottom (PA)
VSC-24 (Vertical Metal/ Concrete Cask)	BNFL Fuel Solutions Corp.	5/7/1993	Palisades (MI) Point Beach (WI) Arkansas Nuclear One (AR)
NAC S/T	NAC International	8/17/1990	Not being used at this time
NAC-C28 S/T	NAC International	8/17/1990	Not being used at this time

Source: "Dry Cask Storage of Spent Nuclear Fuel," Backgrounder, U.S. Nuclear Regulatory Commission, December 2004, pp. 5-6 [http://www.nrc.gov/reading-rm/doc-collections/fact-sheets/dry-cask-storage.html]

High-Level Waste Storage at Defense Related Facilities

Over the years, considerable high-level waste has been generated as a result of the DOE defense activities. Some of the nuclear waste is generated in liquid form during the chemical reprocessing of the spent nuclear fuel to recover uranium or plutonium. Other nuclear materials such as spent nuclear fuel and plutonium are solids. The high-level waste, both in liquid and solid form, is stored on-site in underground tanks at defense facilities at Hanford Reservation, Richland, Washington; Idaho National Engineering and Environmental Laboratory, Idaho Falls, Idaho; and the Savannah River Site, Aiken, South Carolina. (EPA, 2000: 23) Transuranic waste (e.g., clothes, rags, equipment, and sludge contaminated during the weapons production cycle) are transported to the Waste Isolation Pilot Plant (WIPP) in Carlsbad, New Mexico, for disposal.

Advantages/Disadvantages of On-Site Storage

There are a number of advantages and disadvantages associated with storing highly radioactive waste on-site at the point of origin. For example, there are technical, societal, security, and political benefits and costs associated with any radioactive waste storage or disposal solution (see Table 2.4 – Advantages and Disadvantages of On-Site Storage).

Table 2.4 Advantages and Disadvantages of On-Site Storage

Advantages

- minimizes handling of nuclear material
- societal benefit of ensuring equity since the group that receives the benefit of the energy generation is the group that bears the costs associated with the storage
- continuing on-site storage removes the need to transport large quantities of highly radioactive waste across the country
- political advantage of continuing a relatively non-controversial policy that has generated little public opposition

Disadvantages

- intense heat will degrade the storage canisters over time and increase chances for radiation leaks
- dry cask storage system is expensive and added costs are borne by the ratepayers who already help to finance Nuclear Waste Fund (essentially double payment)
- homeland security (e.g., terrorism) and energy security concerns (premature shutdown of commercial nuclear power reactors)

Advantages

Since on-site radioactive waste storage has been safely managed for a number of years, generally there is confidence in that type of storage method. There are a number of advantages associated with the on-site storage method. For example, from a technical standpoint, storing the nuclear waste on-site at the point of origin minimizes the handling of the nuclear material. This reduces the potential for an accident as well as minimizes the risk of radiation exposure to plant workers. However, since the dry cask storage system was not intended to substitute for disposal, eventually the nuclear waste will have to be transferred to another storage container, or be transported to either a centralized storage facility a permanent repository.

Second, there also are societal benefits such as the equity associated with keeping the highly radioactive waste on-site. In essence, the group that receives the economic and energy benefits of the nuclear power that is produced is the same group that bears the safety and economic costs of storing the highly radioactive waste that was generated. Moreover, generally there is a greater acceptance of on-site nuclear waste storage by communities with neighboring nuclear power facilities. Their prior association with nuclear technology likely provides a greater tolerance for nuclear related activities.

Third, continuing on-site storage removes the need to transport large quantities of highly radioactive waste across the country, sometimes through or near large metropolitan areas. This eliminates the risk for either a transportation accident, or a terrorist attack on the nuclear waste shipment and the possible release of radioactive material.

Finally, there are some political advantages of continuing on-site storage. First, continuing the policy of storing highly radioactive waste on-site pressures DOE to find a permanent disposal solution. Second, the commercial nuclear power utilities have increased pressure on DOE to fulfill the federal commitment to take title to the highly radioactive waste. Third, public opposition associated with the current on-site storage regime generally has been muted. Thus, continuing a relatively non-controversial status quo policy (especially one that has been in place for a number of years) involves minimal short-term political risk. Policymakers generally prefer to follow existing policy, or making only minor policy adjustments (i.e., incremental policymaking). Nevertheless, there has been increased litigation in recent years, and conflict is expected to increase in the future as some commercial nuclear power facilities will have to shut down their reactor operations due to a lack of sufficient on-site storage space for the spent nuclear fuel.

Disadvantages

There also are a number of disadvantages associated with the continued on-site storage of highly radioactive waste. First, there are technical considerations that weigh against long-term on-site storage. For example, defective welds and cracked seals have been found on some casks; and explosions and fires associated with the casks have occurred at some locations. Thus, the longer the dry casks remain on-site, the more likely that the intense heat and radioactivity associated with the decaying spent fuel rods will degrade the container material of the dry cask. This could pose a greater safety problem for workers, especially during preparation for the transfer of the spent fuel rods to a transportation cask. (IAEA, 2003: 5)

Second, there are some societal costs associated with on-site dry cask storage. The dry cask storage system is expensive and the added storage costs eventually are borne by the ratepayers. Ratepayers already have paid into the Nuclear Waste Fund to finance the construction and operation of a permanent repository for commercial spent nuclear fuel. Making them continue to fund on-site storage of spent fuel forces essentially makes them pay twice for the nuclear waste storage.

Finally, there also are homeland security and energy security costs associated with the continued storage of highly radioactive waste on-site at numerous locations across the country. First, the tragic events of 9/11 have heightened concerns over the vulnerability of the nuclear waste to terrorism. The on-site waste, especially the dry cask containers, is vulnerable to theft and sabotage. Second, there are energy security concerns since the lack of sufficient on-site storage space for spent fuel could force the premature closure of some commercial nuclear power plants. Since nuclear generated power accounts for about one-fifth of the total U.S. electricity generating capacity, any loss in nuclear power would increase U.S. dependence on foreign sources of energy. For these reasons, the on-site dry cask storage option has become increasingly controversial, which has led to increased political conflict.

Summary

Over the years, commercial nuclear power production and nuclear weapons related defense activities have generated significant quantities of highly radioactive waste. This nuclear waste is being stored across the country at commercial nuclear power facilities and defense plants. Spent nuclear fuel is being stored on-site either in spent fuel pools or in dry cask storage containers and the high-level waste from nuclear weapons production, nuclear naval ships, and defense research activities is being stored on-site at defense plants.

The inability of policymakers to find a long-term solution to the problem of what to do with the waste has generated increased concern over the efficacy of storing large quantities of highly radioactive waste at numerous locations across the country. Clearly, there are legitimate homeland security and energy security concerns associated with the current on-site storage regime. Moreover, these security concerns will increase as the volume of on-site nuclear waste continues to mount. The vulnerability of the waste to a possible terrorist attack, sabotage and the potential for theft of the radioactive material highlight specific homeland security deficiencies associated with storing significant amounts of highly radioactive waste at numerous locations around the country. Thus, homeland security associated with on-site storage would be assessed as poor. Moreover, as on-site storage of spent nuclear fuel reaches capacity at commercial nuclear power facilities, reactors may have to shut down prematurely due to a lack of sufficient on-site storage space. Since nuclear power generated electricity accounts for about one-fifth of the total U.S. electric power generation, any curtailment of production will adversely affect U.S. energy security; thus, energy security would be assessed as poor. Since the current on-site storage regime does not address fundamental homeland and energy security concerns, it presents an unacceptable security risk.

The political reality is that the on-site storage of highly radioactive waste is the preferred political alternative. While conflict over on-site storage has been relatively low, there has been some resistance to expanding on-site dry cask storage (e.g., Minnesota has limited dry cask storage and California has prohibited building additional nuclear power facilities until a pathway for spent nuclear fuel is built). As on-site nuclear waste storage becomes more limited, conflict is expected to increase in the near future as homeland security and energy security concerns become more apparent. In the final analysis, while on-site storage of nuclear waste has been relatively non-controversial and has experienced only isolated conflict, it does not address pressing homeland security and energy security concerns (see Figure 2.6 – On-Site Radioactive Waste Storage: Homeland Security and Energy Security Risk v. Political Risk for a summary of the overall risk analysis associated with on-site storage of highly radioactive waste).

	Poor	*Fair*	*Good*	*Excellent*
Homeland Security Risk	X			
Energy Security Risk	X			
	Low	*Moderate*	*High*	
Political Risk	X*			

Figure 2.6 On-Site Radioactive Waste Storage: *Homeland Security and Energy Security Risk v Political Risk*

* Denotes that political risk likely will increase as on-site storage space becomes limited.

The realization that on-site storage does not provide a long-term answer to the problem of highly radioactive waste storage or disposal has intensified the search for an alternate solution. Chapter 3 examines the centralized interim storage option by providing a historical overview of interim storage, outlines current interim storage proposals, and discusses the advantages and the disadvantages associated with centralized interim storage.

Key Terms

spent nuclear fuel
high-level waste

highly radioactive waste
transuranic waste
Price-Anderson Act
on-site storage
spent fuel pools
re-racking
rod consolidation
dry casks
site-specific license
general license

Useful Web Sources

International Atomic Energy Agency [http://www.iaea.org/]
Nuclear Energy Institute [http://www.nei.org/]
U.S. Department of Energy [http://www.doe.gov/]
U.S. Department of Energy, Office of Civilian Radioactive Waste Management [http://www.ocrwm.doe.gov/]
U.S. Department of Energy, Office of Environmental Management [http://www.em.doe.gov/]
U.S. Environmental Protection Agency [http://www.epa.gov/]
U.S. Nuclear Regulatory Commission [http://www.nrc.gov/]

References

(Callan, 1998) Callan, Joseph L. "NRC's Report to Congress on the Price-Anderson Act," U.S. Nuclear Regulatory Commission, SECY-98-160, 2 July 1998 [http://www.nrc.gov/reading-rm/doc-collections/commission/secys/1998/secy1998-160/1998-160scy.html]

(EIA, 2001) "World Cumulative Spent Fuel Projections by Region and Country," Energy Information Agency, U.S. Department of Energy, 1 May 2001 [http://www.eia.doe.gov/cneaf/nuclear/page/forecast/cumfuel.html]

(EM, 2004) "Environmental Management…Making Accelerated Cleanup a Reality," U.S. Department of Energy, Office of Environmental Management [http://www.em.doe.gov]

(EPA, 2000) "Radiation Protection at EPA: The First 30 Years…Protecting People and the Environment," Office of Radiation and Indoor Air, U.S. Environmental Protection Agency, EPA 402-B-00-001, August 2000 [http://www.epa.gov/radiation/docs/402-b-00-001.pdf]

(Flint, 2000) Flint, Lawrence. "Shaping Nuclear Waste Policy at the Juncture of Federal and State Law," *Boston College Environmental Affairs Law Review*, vol. 28, No. 1, 2000 [http://www.bc.edu/schools/law/lawreviews/meta-elements/journals/bcealr/28_1/05_FMS.htm]

(IAEA, 2003) "The Long Term Storage of Radioactive Waste: Safety and Sustainability: A Position of International Experts," International Atomic Energy

Agency, June 2003 [http://www-pub.iaea.org/MT/publications/PDF/LTS-RW_web.pdf]

(NAS, 2001) "Disposition of High-Level Waste and Spent Nuclear Fuel: The Continuing Societal and Technical Challenges," Committee on Disposition of High-Level Radioactive Waste Through Geological Isolation, Board on Radioactive Waste Management, Division on Earth and Life Studies, National Academy of Sciences, National Research Council, National Academy Press, Washington D.C., 2001 [http://www.nap.edu/openbook/0309073170/html/R1.html]

(NEI, 2004) "Used Nuclear Fuel Management," Nuclear Energy Institute, October 2004 [http://www.nei.org/doc.asp?catnum=3&catid=300&docid=&format=print]

(NRC, 2003a) "High-Level Waste," U.S. Nuclear Regulatory Commission, Revised June 23, 2003 [http://www.nrc.gov/waste/high-level-waste.html]

(NRC, 2003b) "Nuclear Fuel Pool Capacity," U.S. Nuclear Regulatory Commission, Revised 23 June 2003 [http://www.nrc.gov/waste/spent-fuel-storage/nuc-fuel-pool.html]

(NRC, 2003c) "Spent Fuel Pools," U.S. Nuclear Regulatory Commission, Revised 23 June 2003 [http://www.nrc.gov/waste/spent-fuel-storage/pools.html]

(NRC, 2003d) "Locations of Independent Spent Fuel Storage Installations, U.S. Nuclear Regulatory Commission, Revised September 25, 2003 [http://www.nrc.gov/waste/spent-fuel-storage/locations.html]

(NRC, 2004) "Dry Cask Storage of Spent Nuclear Fuel," Backgrounder, Office of Public Affairs, U.S. Nuclear Regulatory Commission, December 2004 [http://www.nrc.gov/reading-rm/doc-collections/fact-sheets/dry-cask-storage.html]

(OCRWM, 1995) "Science, Society and America's Nuclear Waste," Background Notes, Office of Civilian Radioactive Waste Management, U.S. Department of Energy [http://www.ocrwm.doe.gov/pm/program_docs/curriculum/unit_1_toc/14a.pdf]

(OCRWM, 2004) "OCRWM Program Briefing," Office of Civilian Radioactive Waste Management, U.S. Department of Energy [http://www.ocrwm.doe.gov/pm/programbrief/briefing.htm]

(ONEST, n.d.) "Nuclear Energy Timeline and History of the Uranium Program," Office of Nuclear Energy, Science and Technology, U.S. Department of Energy [http://www.ne.doe.gov/]

(Raddatz and Waters, 1996) Raddatz, M.G. and M.D. Waters. "Information Handbook on Independent Spent Fuel Storage Installations," Spent Fuel Project Office, Office of Nuclear Material Safety and Safeguards, U.S. Nuclear Regulatory Commission, NUREG-1571, December 1996.

Chapter 3

Centralized Interim Storage

The delay in establishing a coherent long-term national strategy for spent nuclear fuel storage and disposal has led to a de facto policy of storing the waste on-site. However, due to potential homeland security and energy security implications associated with on-site storage, there has been a search for alternative solutions. Since forging a consensus for a long-term policy for radioactive waste storage has been slow and difficult, the search for a viable interim solution has intensified.

This chapter discusses the politics associated with the interim storage of highly radioactive waste. First, an overview of what countries are successfully operating centralized interim storage facilities is presented. Second, the evolution of interim radioactive waste storage policy is outlined in order to give the reader some historical perspective on the issue. Third, the Skull Valley (Utah) and Owl Creek (Wyoming) private interim radioactive waste storage initiatives are discussed. Finally, the advantages and disadvantages of centralized interim radioactive waste storage are examined.

Overview

Most countries in the world store spent nuclear fuel on-site at the point of origin either in cooling pools or dry cask containers.[1] However, some European countries operate centralized facilities for the interim storage of highly radioactive waste. While Spain currently stores its spent nuclear fuel on-site, it expects to have centralized interim radioactive waste storage facility completed around 2010. (OCRWM, 2001a: 1; MITYC, 2005: 7) Sweden stores it spent nuclear fuel for one year in a cooling pool, and then transports to the Oskarshamn Central Interim Storage Facility for Spent Nuclear Fuel (CLAB) located in southern Sweden. (OCRWM, 2001b: 1; MSDS, 2005: 24) Finally, Switzerland stores it spent nuclear fuel for 1-10 years on-site in cooling pools, and then transports the nuclear waste to a centralized storage facility at Würenlingen. (OCRWM, 2001c: 1; HSK, 2005: 5) The United States does not yet have a centralized interim storage facility for spent fuel and high-level radioactive waste.

Prior to 9/11, safety concerns in the United States about storing spent nuclear fuel in cooling pools and in dry casks were limited. While opposition to dry casks surfaced in some states (e.g., Prairie Island, MN), it was generally muted. For example, in 1995, an open public forum on using dry cask storage at Arkansas

---

1   For the countries that reprocess spent fuel, the vitrified waste from the reprocessing cycle is stored on-site at the reprocessing facility (e.g., Belgium, France, Japan and the U.K.).

Nuclear One was attended by only two individuals other than company officials—both of whom were reporters (Freking, 1995: 1) However, after 9/11, concern over the vulnerability of the nuclear waste stored on-site prompted calls for transporting the waste to a more secure centralized facility. In 2002, Secretary of Energy Spencer Abraham, specifically cited concerns over the vulnerability of the radioactive waste as a major reason to move forward on the Yucca Mountain project (Abraham, 2002a: 3). Abraham clearly was concerned that highly radioactive waste accumulating at nuclear facilities across the country was a major homeland security problem. This perceived vulnerability contributed to the Bush Administration's decision in 2002 (with subsequent congressional approval) to proceed with the Yucca Mountain project to centralize the highly radioactive waste at a more secure facility.

While opponents of a centralized disposal or storage facility acknowledge that maintaining adequate security for the on-site radioactive waste may be a valid concern, they argue that spent fuel rods removed from the reactor will have to remain on-site in cooling pools for some time before the radioactivity levels decrease to the point that they can be transported safely to a centralized facility. Thus, they assert that since there always will be radioactive waste on-site as long as the facility is generating nuclear power, then the nuclear power facility always will be vulnerable to a potential terrorist attack whether or not a centralized facility is built.

Energy security also has been cited as a reason for the need to establish a centralized radioactive waste storage or disposal facility. Secretary of Energy Spencer Abraham noted that energy security was an important factor in his recommendation to the President to proceed with the national highly radioactive waste repository at Yucca Mountain. Clearly, maintaining current levels of energy production would be difficult if commercial nuclear power reactors are forced to shut down prematurely due to a lack of adequate on-site storage space for the radioactive waste. In testimony before Congress, Abraham articulated this position when he stated that the "transportation of nuclear materials is not a function of a repository at Yucca Mountain, but rather is a necessary consequence of the material that continues to accumulate at the 131 sites in 39 States that are running out of room for it." (Abraham, 2000b: 3).

History of Centralized Interim Storage

The concept of a centralized interim storage facility itself has been controversial; and thus, the search for an interim radioactive waste storage solution also has been slow, difficult and fraught with conflict (see Table 3.1 – History of Interim Radioactive Waste Storage).

Opponents of an interim nuclear waste storage facility have voiced concern that any interim facility could become a de facto permanent repository for highly radioactive waste. They are concerned that once an interim facility is established, political realities might preclude building a permanent repository. They have argued that the inability to craft even an interim solution is evidence of the lack of resolve on the part of Congress to solve the problem of highly radioactive waste disposal. (Jacob, 1990: 134)

Table 3.1 History of Interim Radioactive Waste Storage

1950s/1960s–Spent nuclear fuel stored on-site in spent fuel cooling pools.

1972–First centralized interim storage method proposed (RSSF concept).

1982–NWPA establishes Interim Storage Program. Also required DOE to study the Monitored Retrievable Storage (MRS) concept.

1985–DOE proposes sites in Tennessee for possible interim nuclear waste storage facility. State protests, plan dropped.

1986–First license granted by NRC for on-site dry cask storage.

1987–NWPAA authorizes DOE to construct and operate a centralized MRS facility. Also created the MRS Commission, Office of the Nuclear Waste Negotiator, and the Nuclear Waste Technical Review Board. Voluntary siting approach for interim nuclear waste storage created.

1989–MRS Commission recommends two interim storage facilities at DOE sites. One site would handle waste in the event of emergency shutdown of nuclear reactor. The other would accept spent nuclear fuel from decommissioned reactors.

1990s–Voluntary siting approach implemented by Office of the Nuclear Waste Negotiator. Study-grant program established to entice communities to explore possibility of constructing centralized interim storage facility. Four counties submitted applications; only three were approved. Opposition from states prompts federal government to invite tribal governments to investigate siting centralized interim storage facility on tribal lands. Initially, 24 tribes applied for study grants, only two eventually moved on to final study phase. Congress cancels study-grant program in 1993. Authorization for Office of the Nuclear Waste Negotiator expires in 1994.

1992–Consortium of eight utilities formed Private Fuel Storage, LLC (PFS) to begin search for a suitable centralized interim storage facility.

1994–Private Fuel Storage (PFS) begins negotiations with Native American tribes to build a centralized interim storage facility.

1995–Wyoming passes legislation that establishes an application procedure for constructing and operating centralized interim storage facility.

1996–Nuclear Waste Technical Review Board states there is no compelling technical or safety reason to move spent fuel to centralized storage facility for next few years, but centralized facility would be needed by 2010 if permanent repository not available.

1997–PFS enters into lease agreement with Native American Skull Valley Goshute Band (UT) to construct and operate centralized interim storage facility. Owl Creek Energy Project proposes building centralized storage facility in Freemont County (WY).

1997/1998–Congressional attempts to allow DOE to construct an interim radioactive waste storage facility blocked by the Clinton Administration.

1998–DOE misses deadline to take title to spent nuclear fuel being stored at commercial nuclear reactors.

2000–Congressional attempts to pass legislation to have DOE take title to nuclear waste until permanent repository available fails due to inability to override Clinton veto. NRC gives preliminary approval for Skull Valley project.

2003–Atomic Safety Licensing Board (ASLB) rules it cannot recommend license for Skull Valley project due to possible aircraft crash concerns. PFS challenges ruling.

2005–Skull Valley receives ASLB approval for project. Owl Creek project dormant.

2006–Skull Valley lease disapproved by BIA and right-of-way denied by BLM.

The 1970s

In 1972, Atomic Energy Commission Chairman James Schlesinger, presented the first serious proposal for interim radioactive waste storage after it was determined that a proposed permanent disposal facility in the salt beds near Lyons, Kansas was unsuitable.

Schlesinger outlined the *Retrievable Surface Storage Facility (RSSF) concept* as a temporary interim storage measure while attempts to find a more suitable permanent repository continued. Basically, the RSSF concept envisioned a centralized storage facility on the surface where the radioactive waste canisters could be retrieved easily for transfer to a permanent repository. Opponents of this plan, citing concern that the RSSF concept may become a de facto permanent repository, were able to block the proposal. (Bunn, 2001: 7) However, the RSSF concept laid the foundation for the further investigation of interim radioactive waste storage since it was an attractive alternative for states that had nuclear reactors where spent nuclear fuel continued to accumulate.

The 1980s

A decade later, the 1982 Nuclear Waste Policy Act (NWPA) attempted to resolve the inherent tensions associated with the limited space available in the traditional cooling pools by developing new methods for radioactive waste disposal. Although the 1982 Act stated that the commercial nuclear power industry had the primary responsibility for the interim storage of nuclear waste (by maximizing on-site storage), it also gave the federal government the power to establish the *Interim Storage Program* (a federally owned and operated system for the interim storage of spent nuclear fuel) at one or more federally owned facilities. (NWPA, 1982: Sect. 131) The Act also created the Interim Storage Fund that was to be financed primarily by the commercial nuclear power industry through assessed fees for the interim storage. (NWPA, 1982: Sect. 136)

The 1982 Act also required the DOE to begin to study the *Monitored Retrievable Storage (MRS) concept* where radioactive waste would be stored in special canisters above ground until ready for shipment to a permanent repository. An important provision of the Act was the restriction that an MRS facility could not be sited in any state being considered for a permanent repository. Clearly, the enactment of this proposal was motivated by political considerations. First, this would ensure a sharing of the burden of coping with radioactive waste. Second, separating the two concepts would help to soften opposition to the permanent repository. However, the 1982 Act did not specifically address three key considerations concerning the concept of the MRS facility. First, would the MRS facility be short-term interim storage for waste en route to a permanent repository? Second, would the MRS concept be able to provide for the required number of years that would be needed for the lengthy process of devising a permanent solution for radioactive waste disposal? Finally, and most importantly, could the MRS facility become a de facto repository in the event that a permanent site elsewhere did not materialize? (Bunn, 2001: 48)

By the mid-1980s, the inability to respond adequately to these concerns generated opposition to siting a centralized MRS facility. The Department of Energy selected a federal facility, the Clinch River Breeder Reactor site in Oak Ridge, Tennessee, to provide for the interim storage of some radioactive waste. In addition, DOE identified two other Tennessee sites as alternatives to the Oak Ridge location. In spite of the fact that the city of Oak Ridge announced that it would be willing to accept an MRS facility, and Tennessee had a prior history with nuclear technology, the state voiced its opposition and filed a lawsuit opposing the DOE initiative. Subsequently, Tennessee exercised its right to file a *notice of disapproval* (i.e., a "veto" by the state that must be overridden by Congress) as the 1982 Nuclear Waste Policy Act allowed. Congress never took action to override Tennessee's opposition since the 1987 Nuclear Waste Policy Amendments Act revoked the Oak Ridge MRS siting proposal. (Bunn, 2001: 48; Holt, 1998: 6-7)

The 1987 Act authorized the DOE to construct and operate a centralized MRS facility. However, the restrictions placed on siting an MRS facility effectively precluded its construction. For example, Sections 144-48 of the NWPAA impose the following limitations:

- DOE cannot search for a site until a special commission reports on the need for an MRS facility (NWPA Section 144)
- DOE cannot select an MRS site until the Energy Secretary recommends a repository site for presidential approval (NWPA Section 145(b))
- no site in Nevada may be selected for the MRS facility (NWPA Section 145(g))
- construction of an MRS facility cannot begin until the NRC has licensed the construction of a permanent repository (NWPA Section 148(d)(1), and MRS construction must halt whenever the repository license is revoked or construction of the repository ceases (NWPA Section 148(d)(2))
- no more than 10,000 metric tons of spent fuel may be stored at the MRS facility until waste is shipped to a permanent repository (NWPA Section 148(d)(3)), with a limit of 15,000 metric tons after that (NWPA Section 148(d)(4)). (Holt, 1998: 7)

These restrictions made it difficult for the federal government to proceed with the MRS concept. However, some believed that the political opposition "would ease after the first MRS facility went into service." (Carter, 1987: 202)

The 1987 Nuclear Waste Policy Act Amendments also created several new entities that would provide additional oversight over the interim storage of nuclear waste: the MRS Commission, the Office of the Nuclear Waste Negotiator and the Nuclear Waste Technical Review Board. The *MRS Commission* was created to oversee implementation of the MRS concept and report on the viability of an interim radioactive waste storage facility. The *Office of the Nuclear Waste Negotiator* was charged primarily with promoting potential sites for interim storage. The *Nuclear Waste Technical Review Board* was created "to review the technical and scientific validity of DOE's activities associated with investigating the site and packaging and transporting wastes." (GAO, 2001: 5)

In 1989, the MRS Commission recommended that Congress authorize two interim storage facilities at DOE sites. One site would be a limited capacity emergency storage facility (2000 metric tons) for storing spent fuel in the event a nuclear reactor had to be shut down due to an emergency. The other would be a larger facility (5000 metric tons) to accommodate spent nuclear fuel from decommissioned facilities and for operational reactors that would be forced to shut down due to limited on-site storage space. (Holt, 1998: 9; Bunn, 2001: 49) However, there was little support for the MRS Commission proposal due to the continued delay in establishing a permanent repository.

In an effort to reduce opposition to establishing a federally sponsored interim storage facility, the MRS Commission also developed what is known as the "voluntary siting process" which was to be implemented by the Office of the Nuclear Waste Negotiator. Two fundamental premises of the voluntary process were that: (1) any inquiries and preliminary discussions would not require a commitment to proceed with siting a MRS facility; and (2) interested parties could withdraw their application at any stage prior to the agreement being submitted to Congress for approval. The voluntary siting process offered monetary incentives to any party willing to host an interim radioactive waste facility. For example, study grants were available in three phases. Phase I offered a $100,000 grant for communities to learn more about siting a nuclear waste facility as well as to gauge public sentiment to hosting a possible interim storage site. Phase II-A was a $200,000 grant for a more in-depth study of probable sites and local sentiment. The final phase (Phase II-B) would provide over two-million dollars to allow for a more rigorous study of the proposed site. (Gowda, 1998: 231-4; Bunn, 2001: 49) These grants were designed to promote participation in the voluntary siting process, and came with the reassurance that no strings were attached. However, in spite of the monetary incentives and assurances that a party could withdraw from the process at any time without prejudice, there was minimal interest displayed in hosting an interim nuclear waste facility. Only three counties were approved for Phase I grants (Fremont County, Wyoming; Grant County, North Dakota; and San Juan County, Utah), but none of these counties proceeded beyond Phase I. Two of the county applications (Fremont and San Juan) eventually were blocked by the respective governors, and the elected officials responsible for the Phase I grant application in Grant County were removed from office by a recall election. (Bunn, 2001: 49; Gowda, 1998: 233) Thus, if a community did exhibit interest, the state eventually would override the decision by exercising its right to veto the application. As Gowda points out:

> The lack of receptivity on the part of the nation's governors severely compromised whatever hopes for success might have been associated with the Negotiator's voluntary siting process. Not only were the governors unwilling to enter into any communication with the Negotiator, they also thwarted any meaningful participation on the part of those counties that expressed even a preliminary interest in hosting a MRS facility. (1998: 234)

Thus, efforts to site an interim radioactive waste facility with local governments stalled.

The 1990s

Due to opposition by governors to any locally sponsored interim radioactive waste storage initiative, the stage was set for the federal government to begin negotiations with Native American tribes. In all, over 20 different Native American tribes have been involved at one stage or another in the voluntary siting process established by the MRS Commission. While nine tribes moved to the Phase II-A stage of the voluntary siting process, few tribes committed themselves to the Phase II-B stage of the study grant process. (Gowda, 1998: 234-5) The use of Native American tribal lands is an interesting choice with considerable political overtones. The main concern of state officials is that the use of tribal lands could circumvent state sovereignty and erode a state's ability to influence the decision making process. In fact, a consortium of utilities that has entered into an agreement with the Skull Valley Native American tribe in Utah to construct and operate a centralized interim radioactive waste storage facility has argued that the state lacks legal standing to block the project because it would violate tribal sovereign immunity. (PFS, 2001a: 4) Indeed, the concept of tribal sovereignty (i.e., recognized as sovereign entities by the federal government) allows Native American tribes to circumvent state control on a number of issues such as gambling and taxes.[2] (Gowda, 1998: 247) While there is considerable debate over the extent of tribal sovereignty with respect to radioactive waste storage, it does appear that the concept itself would be an attractive means to circumvent opposition from state governments.

While the considerable attention paid to Native American tribes could open the federal government to charges of *environmental justice* (i.e., siting undesirable projects in lower income communities that have reduced political influence), environmental justice concerns may not be fair since the federal government initially directed its attention to state and local communities. Only after efforts to conclude an agreement with state and local governments failed, was there a more concerted effort to consult Native American tribes. As Gowda, (1998: 253) points out:

> Given the high level of concern over nuclear waste risks across the U.S., it can be assumed that governors would veto a nuclear waste facility in their states rather than deal with the political fallout. Therefore, while the governor veto provision can be commended as fair, this feature also possibly ensured that the only participants in the MRS siting process would be Native American tribes.

In spite of the overtures to Native American tribes, however, little real progress for finding an interim solution to the storage of nuclear waste was achieved during the 1990s.

Any serious government sponsored attempt to site an interim facility was blocked by the Clinton Administration until the Yucca Mountain project was found to be viable. Moreover, in a 1996 report to Congress, the Nuclear Waste Technical

---

2   See Clarke, 2002 for an excellent discussion of the origin of Native American tribal sovereignty. She asserts that while Native American tribal sovereignty is fluid and open to interpretation, there has been considerable controversy over the actual extent of tribal sovereignty.

Review Board stated that there "was no compelling *technical* or safety reason to move spent fuel to a centralized storage facility *for the next few years*." (NWTRB, 1996: viii, emphasis original) The report went on to state that since no substantial new storage capacity would be needed until 2010, any decisions on interim storage should wait until the suitability of the permanent nuclear waste repository at Yucca Mountain was determined. However, the report did recommend the creation of a centralized interim storage facility by 2010 due to the anticipated need for increased nuclear waste storage (NWTRB, 1996: viii-ix). The commercial nuclear utilities and the states with nuclear reactors were opposed to any additional delays in removing the on-site waste and began to lobby intensely for building a centralized interim radioactive waste storage facility.

In response to the continued pressure to act, in 1997, both the U.S. House of Representatives and U.S. Senate passed legislation (The Nuclear Waste Policy Act of 1997) that would have authorized the DOE to construct an interim radioactive waste storage facility (40,000 metric ton capacity) at the Nevada Test Site (adjacent to Yucca Mountain). However, in a parliamentary maneuver led by Representative Ensign (R-NV), the House of Representatives returned the legislation to the Senate in 1998. The Clinton Administration announced that it would veto the 1997 Nuclear Waste Policy Act since it "would undermine the credibility of the Nation's nuclear waste disposal program by designating a specified site for an interim storage facility before the viability of that site [Yucca Mountain] as a permanent geological repository has been assessed." (EOP, 1997: 1) Supporters of an interim storage facility in Nevada countered that there were no compelling technical or safety issues that would prohibit construction of a repository at Yucca Mountain. Nevertheless, the threatened veto by the Clinton Administration eroded support for the bill since some Democrats were reluctant to challenge the Administration and further damage the president's political power, already compromised by the numerous investigations into alleged improprieties by the President. Thus, opponents of the legislation in the Senate (led by Senator Reid, D-NV) eventually were able to derail the proposed legislation by narrowly avoiding the passage of a cloture motion that would have ended debate and required the Senate to address the issue.[3] (U.S. Congress, 1997; U.S. Congress, 1998) Nevada's aggressive attempt to fight the siting of any radioactive waste storage or disposal facility within the state has precluded any quick resolution of the issue.

In 1999, the DOE proposed to take title to, and assume the management responsibility for, civilian reactor spent fuel. However, both the governors from states with nuclear reactors and most of the commercial nuclear utility industry opposed this initiative due to concerns that the on-site spent fuel storage under DOE auspices would become de facto permanent waste repositories.

---

3   Initially, the Senate passed the Nuclear Waste Policy Act of 1997 with a 65-34 majority, and the House passed similar legislation with a 307-120 majority. The cloture motion was defeated by four votes (56-39-5). (U.S. Congress, 1997; U.S. Congress, 1998)

The New Millennium

In 2000, Congress again attempted to address the impasse by passing The Nuclear Waste Policy Amendments Act. The new legislation would have authorized the DOE to take title to the spent fuel at the reactor sites and transport the nuclear waste to the permanent repository when it was ready to accept waste. However, the Clinton Administration vetoed the legislation and the Senate narrowly failed to override the veto.[4] The Administration received support primarily from senators (representing states with nuclear reactors) who were concerned that their nuclear facilities would become de facto repositories for the spent nuclear fuel.

The DOE concluded its first agreement with a commercial nuclear utility in 2000, "to address the Energy Department's delay in accepting spent fuel from utilities" by taking title to the nuclear waste. (DOE, 2000: 1) This agreement was intended to serve as a framework for subsequent agreements with other utilities, and would allow a utility to reduce its projected charges paid into the Nuclear Waste Fund to reflect costs reasonably incurred for on-site storage as of a result of the delay in opening a permanent geological repository. However, other utilities challenged the agreement in court arguing that the arrangement for reducing current payments to the Nuclear Waste Fund would result in higher future payments to cover the costs associated with permanent disposal. (Bunn, 2001:51-2) In 2002, a U.S. Court of Appeals ruled against the agreement citing that interim storage does not constitute disposal and thus Nuclear Waste Fund monies cannot be used to offset costs. (NEI, 2002: 1)

In 2002, the Bush Administration decision to move forward on the Yucca Mountain project has had both a positive and negative effect on the perception of the need for a centralized interim storage solution. For some, the Administration's announcement of its intention to pursue constructing the permanent repository has lessened the urgency for finding an interim solution. However, for others the reasons cited for moving forward on a permanent repository (i.e., national and energy security concerns) reinforces the need for a near-term interim solution since Yucca Mountain will not be ready to accept spent nuclear fuel for some time. In addition, concerns that an interim centralized radioactive waste storage facility would become a de facto repository are undercut if it appears that Yucca Mountain will be built.

In 2004, the NRC issued a license to a corporation that specializes in environmental matters to construct and operate an independent spent fuel storage facility located adjacent to the Idaho National Engineering and Environmental Laboratory at Idaho Falls, Idaho. The Idaho Spent Fuel Facility will repackage and store spent nuclear fuel now stored at the DOE Idaho national laboratory. (NRC, 2004: 1)

Although there have been additional proposals to create a government sponsored centralized interim storage facility for highly radioactive waste, little real progress has been made. The continued delay by the federal government in accepting spent

---

4   Senate bill S. 1287 (The Nuclear Waste Policy Amendments Act of 2000) was passed (64-34) in February 2000. The House passed S. 1287 without amendment (253-167) in March 2000. However, in May 2000, the Senate failed to override President Clinton's veto (64-35). (U.S. Congress, 2000).

nuclear fuel prompted the commercial nuclear utilities to move forward with plans to construct a privately funded interim radioactive waste storage facility.

Private Interim Radioactive Waste Storage Initiatives

In 1992, the Minnesota State legislature limited the ability of Northern States Power (a commercial nuclear power utility now called Xcel Energy) to expand dry cask storage at its two reactors. (Bunn, 2001: 50) Facing possible closure of the reactors due to limited spent nuclear fuel storage space, Northern States Power organized a consortium of eight utility interests (Private Fuel Storage, LLC) and began to search for a suitable site to build an interim radioactive storage facility.[5]

The *Private Fuel Storage (PFS)* consortium focused its search on parties that participated in the MRS voluntary siting process in the 1990s, reasoning that it would be easier to find a willing participant from groups who had expressed an earlier interest in hosting an interim radioactive waste storage facility.[6] (Martin, 2004) In 1994, PFS began negotiations with the Native American Mescalero Apache tribe in New Mexico (which had committed to the Stage II-B process of the MRS voluntary siting process). Participation by the Mescalero tribe was not well received by New Mexico since the state had already been selected to host the Waste Isolation Pilot Plant—a repository for transuranic defense waste. Ultimately, negotiations with the Mescalero Apaches broke down; and in 1995, the tribe voted against siting an interim radioactive waste storage facility. (Public Citizen, 2001: 6-7) Subsequently, in 1996, PFS began negotiations with the Native American Skull Valley Band of the Goshute tribe in Utah (which earlier also had committed to proceeded to Stage II-B of the MRS voluntary siting process). In addition, "early on in the site consideration process PFS spoke with four individuals from the Fremont County [Wyoming] business community about a possible site in the Owl Creek area" (Fremont County had received a Stage I grant to explore the possibility of siting an MRS facility).[7] (Martin, 2004) However, since the Skull Valley site was more suitable technically,

---

5   Private Fuel Storage, LLC is a consortium of eight utilities (Xcel Energy, Genoa Fuel Tech, American Electric Power, Southern California Edison, Southern Nuclear Company, First Energy, Entergy, and Florida Light and Power) that according to their web site was created to develop and manage a "safe, clean, temporary storage facility for spent nuclear fuel." (PFS, 2001b: 1) The PFS website is available at: http://www.privatefuelstorage.com.

6   This approach should not be too surprising since the federal government used similar reasoning during the selection process for potential host sites for the national highly radioactive waste repository. When the DOE narrowed the search for a permanent repository in the western U.S. to five potential locations (Richton Dome, Mississippi; Yucca Mountain, Nevada; Deaf Smith County, Texas; Davis Canyon, Utah; and Hanford, Washington), each of these states had a prior involvement with nuclear technology.

7   See (NRC, 2001) "Final Environmental Impact Statement for the Construction and Operation of an Independent Spent Fuel Storage Installation on the Reservation of the Skull Valley Band of Goshute Indians and the Related Transportation Facility in Tooele County, Utah," U.S. Nuclear Regulatory Commission, NUREG-1714, vol. 1, December 2001: 7-1 thru 7-5, for a thorough discussion of the site selection process.

and the use of tribal lands provided some protection against state opposition, the Fremont County site was dropped from consideration.

The Skull Valley Initiative

In 1997, PFS entered into a lease agreement with the Native American Skull Valley Band of the Goshute tribe to construct an interim storage facility for spent nuclear fuel. The proposed *Skull Valley centralized interim storage facility* would be located about 45 miles southwest of Salt Lake City and would be approximately 100 acres in size (see Figure 3.1 – Location of Skull Valley). The complex would be able to store (above ground) up to 4,000 dry casks containing about 10 metric tons of spent fuel each (see Figure 3.2 – Artist Concept of Skull Valley).[8] (PFS, 2000: 1)

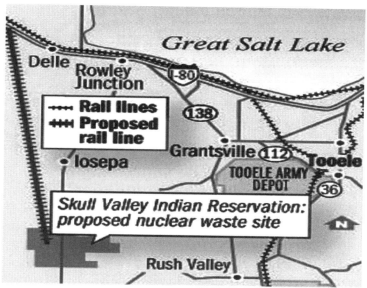

Figure 3.1 Location of Skull Valley
Source: Deseret Morning News, deseretnews.com, Saturday, 9 September 2006 [http://deseretnews.com/dn/print/0,1442,645200115,00.html]

The Approval Process Before the construction and operation of the proposed interim storage facility can begin; four different federal agencies have to approve the plan. First, the NRC would have to issue a license to receive, transfer, and possess the spent nuclear fuel. Second, the Department of Interior's Bureau of Indian Affairs (BIA) would have to approve the lease on tribal lands. Third, the Department of Interior's Bureau of Land Management (BLM) would have to approve the right

---

8   PFS has leased 820 acres of the 18,000-acre reservation. (PFS, 2000:1)   See PFS website available at: http://www.privatefuelstorage.com/project/facility.html.

of way application to construct a new rail spur, as well as an intermodal transfer facility, as required by the Federal Land Policy and Management Act. Finally, the Department of Transportation's Surface Transportation Board (STB) would have to approve the new rail spur to the centralized storage facility. (DOI, 2006: 6-7)

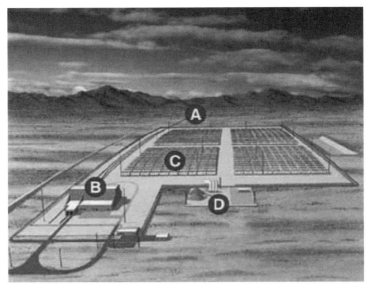

Figure 3.2 Artist Concept of Skull Valley
Legend: A - rail line that will enter the PFS facility, B - cask transfer building, C - dry cask concrete pad, D - concrete plant for storage casks.
Source: Private Fuel Storage, LLC. [http://www.privatefuelstorage.com/project/facility.html]

As required by statute, the tribe submitted the proposed agreement with PFS to the BIA, which approved the arrangement conditional on the successful completion of the environmental impact statement and issuance of a license by the NRC. As required, the final environmental impact statement was completed and PFS subsequently submitted a license application to the NRC. (NRC, 2002: 1) In September 2000, the NRC gave preliminary approval for the project. (Ryan, 2003: 1) However, in March 2003, the NRC's Atomic Safety and Licensing Board (ASLB) ruled that it could not recommend a license until PFS addressed concerns about the possibility of a military aircraft crash at the interim storage facility which is located near an active Air Force base and test range. Initially, PFS proposed operating a smaller facility (336 dry casks instead of 4,000), which would reduce the signature of the area and lessen the probability of an accident. (PFS, 2003a 1-2) However, the ASLB responded that the proposed smaller storage facility did not address the fundamental issue of the probability that an aircraft crash resulting in a possible release of radiation was greater than the one-in-a-million threshold that was established. Subsequently, PFS demonstrated that while the possibility of an aircraft crash may exceed the established threshold, the probability of a release of radiation was still below the established criteria due to the robust nature of the waste storage casks. In response, in February 2005, the ASLB recommended to the NRC that an operating license be granted for the Skull Valley project. (PFS, 2005a: 1) After extensive review, in September 2005, the NRC Commissioners authorized the staff to issue a license to construct and operate a highly radioactive waste storage facility.[9] (NRC, 2005a: 1)

In 2006, two decisions put the proposed centralized storage facility in jeopardy. First, the Bureau of Indian Affairs (BIA) disapproved the proposed lease between PFS and the Skull Valley Goshutes. (DOI, 2006a: 1) Second, the Bureau of Land Management (BLM) refused to grant a "right-of-way" permit to construct and operate a rail line through BLM land to the facility or the construction of an intermodal transfer facility on BLM land. (DOI, 2006b: 1)

In September 2006, The BIA ruled that the earlier conditional lease approval in May 1997 was not valid since a 1991 memorandum from the Assistant Secretary of Indian Affairs specifically prohibited conditional approvals. (DOI, 2006a: 11) Thus, the BIA now considers the conditional approval granted in 1997 to be an expression of intent and not final approval. The recent BIA disapproval of the proposed lease between PFS and the Skull Valley Goshutes was based on a number of factors. First, the increased truck traffic associated with the opening of an adjacent landfill on the reservation in 2004 was not evaluated in the initial environmental impact statement completed in 2001. Second, the impact of the centralized radioactive waste storage facility on the nearby Cedar Mountain Wilderness Area created in 2006 also was not assessed in the initial environmental impact statement. Third, the BIA also cited the recent U.S. 9th Court of Appeals decision which reversed an NRC decision to license the construction and operation of a spent nuclear fuel dry cask storage facility similar to the proposed Skull Valley PFS facility. Fourth, the BIA decision stated that the possible adverse effects of a "terrorist-initiated event" were not adequately considered. Fifth, the BIA cited the BLM decisions not to grant a "right-of-way" for the rail spur across public land to the Reservation or the construction of an intermodal transfer facility as further justification not to grant approval for the project. Sixth, the BIA also cited the potential for the nuclear waste to remain indefinitely if the proposed permanent geological repository at Yucca Mountain is not eventually constructed as an additional reason not to approve the proposed lease. Finally, the BIA expressed concern over the lack or sufficient police protection, as well as no specialized resources to monitor the facility, as further justification for disapproving the project. (DOI, 2006: 20-26)

In September 2006, the BLM decided not to grant a "right-of-way" for either the rail spur or the intermodal transfer facility where the dry casks would be off-loaded from the rail cars to heavy-hauler trucks for shipment to the centralized radioactive waste storage facility. The BLM decision was based on several factors. First, the Bureau stated that it could not grant a "right-of-way" through lands designated as

---

9   One commissioner, who incidentally had worked for Senator Harry Reid from Nevada (NRC, 2005c), registered objections to the NRC decision to authorize the license for the Skull Valley interim storage facility. (NRC, 2005a: 28-32)

the Cedar Mountain Wilderness Area. Second, the Bureau also expressed concern over the adverse impact of increased traffic on the road network. Not only would the weight of the heavy-hauler trucks put a strain on the existing road surface, but also would impact negatively on the traffic flow since the heavily laden trucks with a hazardous cargo must travel at the relatively slow speed of no more than 20 miles per hour. Since the 12 foot wide transport vehicle would have to travel near the center of the road on the parts of the highway that is only 20 feet wide, oncoming traffic would be negatively affected. Third, the BLM also cited the lack of the initial EIS analysis of the impact of the increased traffic flow associated with the adjoining landfill. Fourth, the Bureau stated that storing highly radioactive waste at the intermodal transport facility in preparation for transport to the Skull Valley interim storage facility violates provisions that ban the storage of hazardous materials on BLM lands. (DOI, 2006b: 8-16) Finally, the Bureau stated that the possible construction of the Yucca Mountain permanent geological repository would "reduce, if not eliminate" the need for a centralized interim highly radioactive waste storage facility. (DOI, 2006b: 16)

The BIA and BLM decisions have put the PFS/Skull Valley proposed centralized interim highly radioactive waste storage facility in serious jeopardy. In fact, some accounts are stating that the project is "doomed." (Struglinski, 2006a: 1) Clearly, there are significant challenges ahead for PFS. For example, one of the reasons cited by the BIA for blocking the interim radioactive waste storage initiative was that if the Yucca Mountain repository is not constructed, then the interim radioactive waste storage facility at Skull Valley could become a de facto permanent repository. At the same time, however, the BLM decision cited the progress on Yucca Mountain (i.e., no need for an interim facility) as a reason not to issue a permit for the Skull Valley project. Thus, the two government agencies are citing contradictory reasons with respect to the permanent repository at Yucca Mountain; which in essence rules out an interim highly radioactive waste storage facility whether or not the Yucca Mountain project is ever completed. However, in spite of this latest setback, it is clear that PFS will continue to lobby for the project. In fact, PFS Chairman John Parkyn, has indicated that the consortium will continue to press forward on the Skull Valley project. (Struglinski, 2006b: 1)

Political Ramifications   The Skull Valley initiative has been contentious and politicized from the start. Senator Bob Bennett (R-Utah) acknowledged that the problem is political, not scientific nor technical. (Fahys, 2002b: 1) In an effort to prevent the Skull Valley project from moving forward, Governor Michael Leavitt (R-Utah) issued an executive order creating a multi-agency task force "that will do everything possible to block storage of high level nuclear waste in Utah." (Utah.gov, 1997: 2). In fact, the governor of Utah has vowed to block radioactive waste shipments at the border. (Talhelm, 2005: 2-3) Bolstering its opposition to the proposed interim radioactive waste storage facility, Utah created the Office of Opposition to High Level Nuclear Waste Storage under the Department of Environmental Quality (DEQ) to act as a clearinghouse for opposition to the Skull Valley project. (Utah DEQ, 2000: 1) In 1997, Utah filed a number of "contentions" (i.e., concerns/objections) concerning the proposed nuclear waste storage facility.[10] (NRC, 1997) The objections cover a wide range of issues involving all phases of the proposed operation. Much of Utah's concern has involved the conflict over tribal sovereignty and state's rights.

Utah has attempted to block the proposal at every level. For example, the state has filed a number of lawsuits to block the Skull Valley venture. Utah argued that the NRC does not have the authority to license the facility. However, a U.S. Court of Appeals ruling in 2000 asserted that state's claim was premature and that it would have ample opportunity to raise its concerns at a later date. (*Utah v. PFS*, 2000: 2) In a further blow to Utah's efforts to block the Skull Valley initiative, the NRC ruled in December 2002 that it did have the authority to license a private storage site for spent nuclear fuel. (Fahys, 2002d: 1) Utah also has attempted to block the Skull Valley venture by passing several laws that would impose a substantial burden on the proposed project. Utah passed laws which would: (1) impose a $150 billion up-front bond for the facility; (2) impose a 75 percent tax on anyone providing services to the project; (3) bar Tooele County from providing municipal services such as fire and police; and (4) restrict the transport of nuclear waste into the state. (Roche, 2002: 1) Moreover, in an effort to curry favor with the Native American tribes, Utah passed legislation to study the needs and requirements of Native American reservations within the State. (Utah.gov, 2001: 1) However, in July 2002, a U.S. District Judge ruled that the state laws unfairly hindered the project and prohibited the state from enforcing those policies. (*PFS v. Utah*, 2002: 1; Nuke-Energy, 2002: 1)

Other groups also have voiced their opposition to the Skull Valley project. The "NO Coalition" (a group of citizens and various political, business and former military leaders) has been especially active in its opposition to the interim storage facility and has cited numerous environmental and safety concerns over the proposed project.[11] (Bryson, 2002: 1) Opponents of the Skull Valley initiative have expressed frustration over their inability to derail the proposal. Governor Leavitt echoed that frustration when he said, "We are not a fair match with the federal government and utilities." (Fahys, 2002a: 1) Opponents also have attempted to block the venture

---

10 See, (Utah DEQ, 1997) "State of Utah's Contentions on the Construction and Operating License Application by Private Fuel Storage, LLC for an Independent Spent Fuel Storage Facility," available at: http://www.deq.utah.gov/Issues/no_high_level_waste//documents/11_23_97.PDF. Also see, (Utah DEQ, 2002) "Information on Private Fuel Storage's Proposal to Locate a High Level Nuclear Waste Storage Facility on the Skull Valley Goshute Indian Reservation" for a summary of a number of contentions, available at: http://www.deq.utah.gov/Issues/no_high_level_waste/concerns.htm. Utah also submitted over 40 procedural and general comments in opposition to the Draft Environmental Impact Statement. See, "Comments Submitted by the State of Utah, September 20, 2000, on the Draft Environmental Impact Statement," available at: http://www.deq.utah.gov/Issues/no_high_level_waste/commentsindex.htm.

11 For a detailed explanation of the Coalition's concerns, see NO Coalition (2000), "White Paper Regarding Opposition to the High-Level Nuclear Waste Storage Facility Proposed by Private Fuel Storage on the Skull Valley Band of Goshute Indian Reservation, Skull Valley, Utah," The Coalition Opposed to High-Level Nuclear Waste, November 28, 2000.

by linking it with the Yucca Mountain project. They argue that if the Yucca Mountain project is not approved by the federal government, then the Skull Valley facility could ultimately become a de facto repository for highly radioactive waste. However, until the 2006 BIA and BLM decisions, this approach has not proved fruitful for a number of reasons. First, although Yucca Mountain is behind schedule, it has gained some momentum under the Bush Administration. Second, if Yucca Mountain is not approved, then there will be an even greater need for some type of centralized storage facility for spent nuclear fuel since a number of reactors would be in danger of having to shut down due to limited on-site storage space. Third, the increased risk of terrorism against nuclear facilities has prompted increased calls for centralizing storage of highly radioactive waste away from population centers. This view is echoed by PFS spokesperson Sue Martin:

> ...any outcome over Yucca Mountain is a win-win situation for the PFS project. If Congress decides that Yucca Mountain is suitable, then Utah's contention that the Skull Valley interim storage facility will become a de facto permanent repository would be alleviated. [Conversely] if Congress decides that Yucca Mountain is not suitable, then the need for an interim centralized storage facility may increase [significantly]. (Martin, 2002a)

In fact, opponents of the project have conceded that the decision to move ahead on Yucca Mountain "makes the Private Fuel Storage proposal more probable." (Coyne, 2002: 1) At the same time, PFS has expressed frustration with Utah's attempt to derail the Skull Valley initiative. According to Martin:

> It's frustrating to us, said PFS' Martin, "that the state continues along these lines of defending laws that we believe to be unconstitutional, and to continue to take an over-my-dead-body position, when we would be willing to sit down and begin discussing issues of common interests. (Fahys, 2002c: 1)

While the fate of the Skull Valley project is uncertain, there have been other interesting developments. For example, at the same time the governor has aggressively attempted to block the initiative, a proposal by several state legislators to build a state operated storage site for highly radioactive waste quietly began to gather some momentum. This initiative, labeled *Plan B*, initially surfaced in late 2000, and received renewed attention in early 2003. Under Plan B, the state would construct and operate a nuclear waste facility on state school trust lands. (Fahys and Harrie, 2003: 1) While support for this initiative has been limited, it does highlight how some Utah lawmakers understand the Skull Valley project could still receive federal approval. Apparently, lawmakers have concluded it would be better to have a nuclear waste storage facility under the control of the State of Utah rather than a facility operated by a private consortium on tribal lands outside of state jurisdiction. Utah Senate Majority Leader Mike Waddoups (R-Taylorsville) reportedly stated: "If it's going to happen, maybe we ought to have some oversight and get some benefits and still help the Goshutes." (Fahys, 2003a: 1) Understandably, "Plan B" has not been well received by PFS. According to Martin, "it appears as though the state is attempting to divert the economic benefit from the Goshutes [into state coffers] by proposing to store spent nuclear fuel on state trust land." (Martin, 2003a) Martin also reportedly stated, "If that's the intent, to rob the Skull Valley [Goshute] Band of economic development they have been working on for 13 years, then that is just shameful." (Fahys, 2003a: 3)

In other related developments, six of the eight PFS members promised Utah's senators that no funds would be committed for the facility as long as the Yucca Mountain project proceeds in a "timely" fashion. However, according to PFS spokesperson Martin, "timely" is "subjective for each utility." (Martin, 2002b) Moreover, it appears that the DOE Secretary Abraham may have promised not to commit federal funds to the Skull Valley facility as a means for securing support from Utah's two senators for the Yucca Mountain project. However, Martin has stated that PFS never intended to use federal funds that were set aside for nuclear waste disposal (i.e., from the Nuclear Waste Fund) for the Skull Valley project and stressed that the project will go forward as planned since "It was always our intent that construction will be financed through service agreements with our customers." (Spangler and Davidson, 2002: 1) Thus, PFS anticipates that the utilities that want to contract for storing their spent nuclear fuel will finance the project. Given that some commercial nuclear utilities may be forced to shut down their reactors within the next few years due to a lack of adequate storage space for spent nuclear fuel, the Skull Valley project is an attractive alternative since the interim storage facility could be ready for operation in the near future. (Ryan, 2003: 1)

Why has PFS chosen to pursue an interim solution, and why Skull Valley? Clearly, economic factors have been the dominant catalyst. The driving force for PFS seeking to construct an interim waste facility has been predicated on the fact that some of the consortium's current reactors would have to be shut down due to a lack of sufficient on-site storage space. Thus, in order to continue current levels of commercial nuclear energy generation, it has become necessary to find an alternative storage method. Because the Yucca Mountain project will not be ready to accept high-level waste for some time (even if eventually approved), an interim solution has become an attractive option. Since the states blocked local initiatives to explore hosting an interim nuclear waste site under the government sponsored voluntary MRS program, the only other alternative was to explore the use of Native American tribal lands since they possessed a certain degree of sovereignty and autonomy from the states. Private Fuel Storage focused its search on tribal entities that actively pursued the MRS voluntary siting program. After negotiations with the Mescalero tribe in New Mexico broke down, the Skull Valley Goshute tribe was approached since they also had participated in the MRS process. The Goshutes also have pursued hosting an interim radioactive waste facility due to economic considerations. Tribal leaders realize that the facility will create jobs and generate significant income from storage fees paid by PFS. While the use of Native American tribal lands has allowed project supporters to circumvent state sovereignty over the process, Utah's strenuous efforts to stop the Skull Valley initiative have delayed opening the project.

Prior to the September 2006 rulings by the Bureau of Indian Affairs and the Bureau of Land Management, it appeared that the PFS/Skull Valley interim radioactive waste facility was moving forward. In fact, the project was projected to be ready as early as 2008. (PFS, 2005b: 2) However, the recent BIA and BLM rulings have seriously undermined the proposal.

The Owl Creek Venture

A group of private business interests in Wyoming proposed in 1997 to build and operate an interim radioactive waste storage facility on private lands for spent nuclear fuel. The *Owl Creek Energy Project* would have sited the facility in Fremont County, Wyoming. The proposed project would have been sited in an area located near the geographic middle of the state adjacent to the town of Shoshoni (population 550), approximately 90 miles east of Casper, Wyoming. The facility design comprised about 100-200 acres, and would have been located within a privately owned section of undeveloped land approximately 2700 acres in size. (Stuart, 1999: 3-4) What made the Owl Creek initiative unique is that it is a wholly private venture.[12] The proposed facility would have had a capacity of 40,000 metric tons of spent nuclear fuel with a design lifetime of 40 years. (OCEP, 2000: 3)

The Approval Process In an effort to streamline the process, the project was modeled after the DOE's Central Interim Storage Facility (CISF) design. The proposed Owl Creek facility design incorporated the following specifications:

- a capacity of 40,000 metric tons of spent fuel
- a design lifetime of 40 years
- an ability to handle a variety of different container systems
- is privately sponsored and operated
- only sealed spent fuel containers will be accepted
- only spent fuel from U.S. utilities will be accepted
- transportation will be via rail only
- a fully funded decommissioning program that would return the site to its original condition. (OECP, 2000: 3; Stuart, 1999: 6)

The Owl Creek project initially got its genesis in the early 1990s during the voluntary siting initiative under supervision of the DOE Office of the Nuclear Waste Negotiator. Fremont County received a Phase 1 grant ($100,000) to learn more about siting a nuclear waste storage facility as well as gauge public sentiment to a possible siting proposal. Although there appeared to be local public support for the project, the state declared its opposition to the proposed facility, which halted further study of the issue. (Gowda, 1998: 5; Stuart, 1999: 3)

Political Ramifications In 1995, Wyoming passed legislation that would permit applications for constructing and operating an interim radioactive waste storage facility. A two-step process was established: (1) conduct a preliminary nonbinding

---

12 See: Stuart, Ivan F. "Owl Creek Energy Project: A Solution to the Spent Fuel Temporary Storage Issue," March 1999, for a detailed description of the various interests involved in the project.

feasibility study (which required the governor's approval) to determine if the proposed site adhered to established environmental and safety criteria; and (2) apply to the Wyoming State Department of Environmental Quality for a permit to construct and operate a storage facility for spent nuclear fuel. (Stuart, 1999: 5) This legislation was shepherded through the Wyoming Legislature by prominent state legislators who, coincidently, were reportedly on corporate boards associated with the Owl Creek Energy Project. (Moore, 2000: 1)

In 1997, the Owl Creek Energy Project announced it intended to apply for approval to construct and operate a private interim radioactive waste storage facility. The following year, an evaluation of the suitability of the site and a projected benefits package outlining the compensation to state and local governments were completed. However, Governor Jim Geringer denied permission for the Owl Creek Energy Project to proceed with the preliminary feasibility agreement and study. (Geringer, 1998) After that, interest in the project diminished; and in 2002, collapsed altogether due to a lack of financial support. (Jones, 2005)

Currently, it appears that the Owl Creek project has little chance for recovery. As a private venture on public property it does not enjoy the protection afforded by Native American tribal lands. However, while there is state-level opposition to the proposed project, there appears to be a surprising degree of local support for the venture. For example, the Riverton Community Development Association and the Shoshoni Chamber of Commerce issued resolutions or letters in support of a feasibility study for the proposed project. Moreover, the Wyoming Mining Association reportedly issued a resolution of support for the Owl Creek project. (Stuart, 1999: 9)

Why did Wyoming voluntarily take steps to permit the possible construction of an interim radioactive waste storage facility within the state, and why was Owl Creek singled out as the possible site? Again, economic considerations prevailed. In 1995, Wyoming passed a law that established an application process for constructing an interim radioactive waste storage facility in the state. Critics of the legislation argue that the Wyoming uranium mining industry was the catalyst behind the proposal since the market for newly mined uranium would be adversely affected if commercial nuclear reactors were forced to be shut down due to the lack of sufficient on-site spent fuel storage space. Key state legislators who have supported the Owl Creek initiative reportedly had ties to the Wyoming uranium mining industry. Because Fremont County had earlier participated in Stage I of the MRS voluntary siting program, it was a logical choice since it appeared there would be sufficient local support for approving a proposed interim radioactive waste facility. Supporters of an interim radioactive waste storage facility realized that the project would generate significant income and jobs for the local area, as well as help ensure a continued need for uranium.[13] In fact, it has been estimated that the over the life of the project, almost $2 billion of local income could be generated. (Stuart, 1999: 8)

---

13 According to Mr. Robert Anderson, President of the Owl Creek Project, the facility would pay a one-time up-front fee of $5/kilogram ($62 million) and an additional $1/kilogram annual storage fee and hazardous waste transportation fee. (trib.com, 2001: 2)

Advantages/Disadvantages of Centralized Interim Storage

There are a number of advantages and disadvantages associated with centralized interim storage. For example, there are homeland security and energy security benefits associated with the centralized interim storage option. However, these benefits would not be long-term. Moreover, there are societal and political costs associated with pursuing a centralized interim storage option (see Table 3.2 – Advantages and Disadvantages of Centralized Interim Storage).

Table 3.2 Advantages and Disadvantages of Centralized Interim Storage

Advantages

- improved homeland security by reducing some of the most vulnerable waste
- improved energy security by providing temporary pathway for on-site waste and avoiding premature shutdown of commercial nuclear reactors

Disadvantages

- would address homeland security and energy only for the near-term
- centralized option would require transport of significant quantities of highly radioactive waste across the country
- centralized interim storage facility would be a high-value target for terrorists
- societal costs associated with disproportionately shifting health and safety risks to group that hosts facility
- increased political conflict with centralized storage option

Advantages

There are both homeland security and energy security advantages associated with centralizing spent nuclear fuel at an interim storage facility. First, a centralized storage option would provide greater protection against a terrorist attack on the spent nuclear fuel stored at numerous sites around the country. Centralizing the waste would reduce the number of sites where vulnerable nuclear waste is being stored and becomes even more critical in the near future as nuclear power plants are decommissioned. Since there have been problems documented with adequate security at some sites in the past, it would appear prudent to consolidate the most vulnerable on-site waste at a centralized storage facility away from any major population centers. Additionally, it is likely that the security at a centralized storage facility should be more robust since it would be considered a high-value target.

Energy security concerns would be partially addressed by providing a temporary pathway for spent nuclear fuel. This would prevent commercial nuclear power reactors from having to shut down their reactors prematurely due to a lack of sufficient on-site storage space. If the nuclear power facility were in danger of reaching its storage capacity for spent nuclear fuel, additional on-site storage space could made available by removing some of the waste already on-site. This consideration becomes even more important if construction of a national repository continues to be delayed, or when constructed, it experiences operational problems that results in either slowing or stopping waste acceptance.

Disadvantages

There also are a number of homeland and energy security disadvantages associated with the centralized interim storage option. First, a centralized storage option only would address homeland security and energy security concerns for the mid-term since it would not have the storage capability to store significant quantities of spent nuclear fuel. Thus, there would continue to be some spent nuclear fuel and high-level waste stored on-site at a number of reactor sites across the country.[14] While this may reduce homeland security concerns, it would not eliminate them. Second, any centralized option would require the transport of highly radioactive waste across great distances, sometimes near heavily populated centers.[15] While the transportation safety record has been good, there is no guarantee that an accident or terrorist attack would not occur. Moreover, opponents argue that it is prudent to wait until a national repository is ready to accept highly radioactive waste so that the nuclear waste will not have to be transported twice. Third, a centralized interim storage facility would be an attractive target for terrorists since a large quantity of highly radioactive waste would be stored above ground. However, a centralized storage facility would be expected to have more robust security.

There also are societal concerns associated with the centralized interim storage option. The societal costs associated with centralized interim storage revolve around the principle of equity. First, the group that receives the benefit of the nuclear power that was produced should be the same group that bears the burden of coping with the radioactive waste that was generated as a result of the energy production. Thus, any centralized option would disproportionately shift the burden to the community that would host the interim storage facility. Second, there are perceived health and safety risks associated with any centralized radioactive waste facility. Clearly, concerns over possible radiation exposure whether due to an accident, terrorist attack, or daily operation of the facility are understandable. Third, the stigma associated with hosting a centralized nuclear waste facility could have a negative economic impact by adversely affecting property values or tourism. Finally, the concept of centralized interim highly radioactive waste storage has been controversial since opponents

---

14 For example, while the two proposals (Skull Valley, UT; and the Owl Creek, WY) have planed to store up to 40,000 metric tons of spent nuclear fuel, it is likely that only the Skull Valley initiative will be successful – at least in the near-term. (PFS, 2000: 1; OECP, 2000: 3)

15 See (Rogers and Kingsley, 2004) Rogers, Kenneth A. and Marvin G. Kingsley. "The Transportation of Highly Radioactive Waste: Implications for Homeland Security," *Journal of Homeland Security and Emergency Management*, vol. 1, Issue 2, Article 13, 2004, [http://www.bepress.com/jhsem/vol1/iss2/13] for a discussion of the risks associated with highly radioactive waste transportation.

have voiced concern that any interim facility could become a de facto repository. Opponents of a centralized interim storage regime are concerned that the creation of any interim storage option will delay creating a permanent solution by removing the incentive to move forward on a permanent repository. This ultimately could defer addressing the problem for the foreseeable future, which would result in future generations being responsible for having to dispose of the nuclear waste. Thus, any interim storage solution has been inextricably tied to progress on permanent disposal. While a centralized interim storage solution will address some of the more immediate problems associated with the continued on-site storage regime, it does not permanently solve the problem and offers only a temporary fix to the problem of what to do with the spent nuclear fuel.

Clearly, centralized interim storage of highly radioactive waste has generated controversy. Those who support a centralized storage option tend to have faith in the technical capability to store the waste safely. In addition they focus on the economic benefit of jobs, disposal fees, and infrastructure improvements such as upgraded utilities and roads. Opponents, on the other hand, tend to focus on the possible health and safety risks of a possible radioactive release. Moreover, there is the question of equity for the host state. For example, Utah does not have any nuclear power facilities, yet it could be the host for the highly lethal waste generated in other states.

Although a centralized storage option was first proposed in the early 1970s, conflict and politicization of the issue have prevented the construction of any centralized facility. Initial efforts by local jurisdictions to consider hosting a centralized interim storage facility have been blocked by the affected state. Thus, while a centralized storage option for highly radioactive waste would partially address both homeland security and energy security concerns, the politicization of the issue has produced considerable political conflict in the states that possibly would host such a facility. Since elected officials tend to avoid contentious issues, it has been easier to follow the on-site storage option since it has generated less conflict than pursuing a centralized storage option.

Summary

Clearly, the interim storage issue is inextricably tied to the Yucca Mountain saga. If Yucca Mountain goes forward as planned, concerns over an interim storage facility becoming a de facto repository will be minimized. On the other hand, if Yucca Mountain does not materialize, then there will be an increased need for an interim storage facility due to the necessity to find additional storage capacity for commercial nuclear energy facilities that are in danger of closing down their reactors due to a lack of adequate on-site storage for spent nuclear fuel.

The fate of the interim storage of highly radioactive waste storage is still uncertain. On the one hand, the realities of the need to continue to produce nuclear energy (coupled with the increased realization that the radioactive waste must be protected from a potential terrorist attack) will prompt efforts to move forward with the centralized storage of spent nuclear fuel. On the other hand, it is clear that the process has become so contentious and politicized, that any siting proposal will meet stiff resistance from the state where the proposed facility will be located.

The federal government ultimately may be more receptive to interim storage proposals in order to avoid the premature shutdown of reactors due to a lack of adequate storage space, which would have a negative impact on energy security. In spite of the fact that the search for a solution to the troublesome problem of what to do with radioactive waste has been slow and arduous, policymakers will soon be forced to come to grips with the radioactive waste storage issue in the near-term rather than the long-term.

The political reality is that while interim highly radioactive waste storage has been contentious at the state-level, national opposition has been somewhat muted. Thus, the political risk is moderate. However, a centralized interim storage facility does not fully address the homeland security and energy security concerns. While homeland security could be improved by removing the most vulnerable on-site spent nuclear fuel, the facility would not have sufficient storage space for all of the current on-site nuclear waste. Since a centralized interim storage option would only be a temporary solution to the most pressing security concerns, its overall impact on homeland security is assessed as fair. Energy security also would be improved since a centralized interim storage facility would allow spent nuclear fuel to be removed from commercial nuclear power facilities that are in danger of having to shut down their reactors due to a lack of sufficient on-site storage space. Thus, at least for the near term, commercial nuclear power production would not have to be curtailed due to a lack of storage capacity for the spent nuclear fuel. However, since this is not a permanent solution, energy security would be assessed only as good (see Figure 3.3 – Interim Radioactive Waste Storage: Homeland Security and Energy Security Risk v. Political Risk for a summary of the overall risk analysis associated with the interim storage of highly radioactive waste).

	Poor	Fair	Good	Excellent
Homeland Security Risk		X		
Energy Security Risk			X	
	Low	Moderate	High	
Political Risk		X		

Figure 3.3 Interim Radioactive Waste Storage: *Homeland Security and Energy Security Risk v Political Risk*

In any case, the search for a solution of how best to cope with the growing stockpiles of highly radioactive waste will continue. Chapter 4 discusses the search for a permanent solution, compares national approaches to highly radioactive waste disposal, provides an overview of the history of establishing a geologic repository in the United States, examines the Yucca Mountain project, and discusses the advantages and the disadvantages associated with a permanent repository.

Key Terms

Retrievable Surface Storage Facility (RSSF) concept
Interim Storage Program
Monitored Retrievable Storage (MRS) concept
notice of disapproval
MRS Commission
Office of the Nuclear Waste Negotiator
Nuclear Waste Technical Review Board
environmental justice
Private Fuel Storage, LLC
Skull Valley centralized interim storage facility
Plan B
Owl Creek Energy Project

Useful Web Sources

Deseret News [http://deseretnews.com]
Nuclear Energy Institute [http://www.nei.org]
Nuclear Waste Policy Act of 1982 (P.L. 97-425) [http://www.ocrwm.doe.gov/documents/nwpa/css/nwpa.htm]
Nuclear Waste Technical Review Board [http://www.nwtrb.gov]
Private Fuel Storage, LLC [http://www.privatefuelstorage.com]
U.S. Congress [http://thomas.loc.gov]
U.S. Department of Energy [http://www.doe.gov]
U.S. Nuclear Regulatory Commission [http://www.nrc.gov]
Utah Department of Environmental Quality [http://www.deq.state.ut.us]

References

(ABC News OnLine, 2004) "Report urges waste dump plan be abandoned," 18 February 2004 [http://www.abc.net.au/news/newsitems/s1047286.htm]
(Abrahms, 2005) Abrahms, Doug. "As Yucca project stalls, Utah nuke waste dump hits fast track, *Reno Gazette Journal*, 4 April 2005 [http://www.rgj.com/news/stories.html/2005/04/02/96157.php]
(Bryson, 2002) Bryson, Amy Joi. "Utah Group just says no to N-waste at Goshute," *Deseret News*, 27 April 2002.
(Bunn, 2001) Bunn, Mathew, et. al. "Interim Storage of Spent Nuclear Fuel: A Safe, Flexible, and Cost Effective Near-Term Approach to Spent Fuel Management,"

A Joint Report from the Harvard University Project on Managing the Atom and the University of Tokyo Project on Sociotechnics of Nuclear Energy, June 2001 [http://ksgnotes1.harvard.edu/BCSIA/Library.nsf/pubs/spentfuel]

(Carter, 1987) Carter, Luther J. *Nuclear Imperatives and Public Trust: Dealing with Radioactive Waste*, Washington D.C.: Resources for the Future.

(Clarke, 2002) Clarke, Tracylee. "An Ideographic Analysis of Native American Sovereignty in the State of Utah: Enabling Denotative Dissonance and Constructing Irreconcilable Conflict," *Wicazo Sa Review*, University of Minnesota Press, vol. 17, No. 2, Fall 2002: 43-63.

(Coyne, 2002) Coyne, Connie. "Utah Site Foes Call Approval of Yucca Waste a Bitter Pill," *The Salt Lake Tribune*, 16 February 2002 [http://www.energy-net. org/IS/EN/NUKE/WST/FED/NEWS/NP-1644.202]

(DOE, 2000) "First Agreement Reached with Utility on Nuclear Waste Acceptance," *Department of Energy News*, U.S. Department of Energy, 20 July 2000 [http:// www.energy.gov/HQPress/releases00/julpr/pr00186.htm]

(DOI, 2006a) "Record of Decision for the Construction and Operation of an Independent Spent Fuel Storage Installation (ISFSI) on the Reservation of the Skull Valley Band of Goshute Indians (band) in Tooele County, Utah," U.S. Department of Interior, Bureau of Indian Affairs, 7 September 2006 [http://www. deq.utah.gov/Issues/no_high_level_waste/ROD%20PFS%2009072006.pdf]

(DOI, 2006b) "Record of Decision Addressing Right-of-Way Applications U 76985 and U 76986 to Transport Spent Nuclear Fuel To the Reservation of the Skull Valley Band of Goshute Indians," U.S. Department of Interior, Bureau of Land Management, 7 September 2006 [http://www.deq.utah.gov/Issues/no_high_level_waste/ROD%20Right%20of%20Way%20Skull%20Valley%2009072006. pdf]

(EOP, 1997) "Statement of Administration Policy: S. 104 – Nuclear Waste Policy Act of 1997," Executive Office of the President, 7 April 1997 [http://clinton3. nara.gov/OMB/legislative/sap/105-1/S104-s.html]

(Fahys, 2002a) Fahys, Judy. "Utah Fears Waste Plan Is Shoo-In," *The Salt Lake Tribune*, 22 April 2002 [http://www.energy-net.org/IS/EN/NUKE/WST/FED/ NEWS/NP-2242.402]

(Fahys, 2002b) Fahys, Judy. "Senator Cites N-Dump Politics," *The Salt Lake Tribune*, 27 April 2002 [http://www.energy-net.org/IS/EN/NUKE/WST/FED/ NEWS/NP-2710.402]

(Fahys, 2002c) Fahys, Judy. "State Files Its Appeal to Halt Goshute Plan," *The Salt Lake Tribune*, 16 August 2002 [http://www.energy-net.org/IS/EN/NUKE/WST/ FED/NEWS/NP-1627.802]

(Fahys, 2002d) Fahys, Judy. "Feds Are No Help to Utah on N-Waste," *The Salt Lake Tribune*, 19 December 2002 [http://www.sltrib.com/2002/dec/12192002/Utah/ 12796.asp]

(Fahys, 2003) Fahys, Judy. "Goshute plan suffers setback," *The Salt Lake Tribune*, 27 September 2003 [http://www.sltribune.com/2003/sep/09272003/utah/96424. asp]

(Fahys and Harrie, 2003) Fahys, Judy and Dan Harrie. "'Plan B' Aims to Outbid Goshutes' N-Waste Site,'" *The Salt Lake Tribune*, 6 February 2003 [http://www.sltrib.com/2003/feb/02062003/Utah/26765.asp]

(FANC, 2006) *Second Meeting of the Contracting Parties to the Joint Convention on the Safety of Spent Fuel Management and on the Safety of Radioactive Waste Management*, Federal Agency for Nuclear Control, Kingdom of Belgium, National Report, May 2006 [http://www.avnuclear.be/avn/JointConv2006.pdf]

(GAO, 2001) "Nuclear Waste: Technical, Schedule, and Cost Uncertainties of the Yucca Mountain Project," General Accounting Office, Report to Congressional Requesters (Senator Harry Reid), GAO-02-19, December 2001.

(Geringer, 1998) Geringer, Jim. "Geringer Says No to Owl Creek Nuclear Waste Project," Office of the Governor, State of Wyoming, 6 April 1998.

(Gowda, 1998) Gowda, M.V. Rajeev and Doug Easterling. "Nuclear Waste and Native America: The MRS Siting Exercise," *Risk Health, Safety and Environment*, Summer 1998: 229-58.

(Holt, 1998) Holt, Mark. "Civilian Nuclear Spent Fuel Temporary Storage Options," Congressional Research Service, 96-212 ENR (Updated 27 March 1998) [http://cnie.org/NLE/CRSreports/Waste/waste-20.cfm]

(HSK, 2005) *Implementation of the Obligations of the Joint Convention on the Safety of Spent Fuel Management and on the Safety of Radioactive Waste Management*, Second National Report of Switzerland in Accordance with Article 32 of the Convention, Department of Environment, Transport, Energy and Communications, September 2005 [http://www.hsk.ch/english/files/pdf/joint-convention_CH_sept-05.pdf]

(IEER, 1999) "International Repository Programs," Institute for Energy and Environmental Research, May 1999 [http://ieer.org/sdafiles/vol_7/7-3/repos.html]

(Jacob, 1990) Jacob, Gerald. *Site Unseen: The Politics of Siting a Nuclear Waste Repository*, Pittsburgh, PA: University of Pittsburgh Press, 1990.

(Jones, 2005) Jones, Tom A. Former lobbyist for Owl Creek Energy Project, Telephone Interview, 29 March 2005.

(Katz, 2001) Katz, Jonathan L. "A Web of Interests: Stalemate on the Disposal of Spent Nuclear Fuel," Policy Studies Journal, vol. 29, Issue 3, 2001: 456-77.

(Macfarlane, 2001a) Macfarlane, Allison. "Interim Storage of Spent Fuel in the United States," *Annual Review of Energy and the Environment*, vol. 26, 2001: 201-35.

(Macfarlane, 2001b) Macfarlane, Allison. "The Problem of Used Nuclear Fuel: Lessons for Interim Solutions From a Comparative Cost Analysis," *Energy Policy*, vol. 29, 2001: 1379-89.

(Martin, 2002a) Martin, Sue. Spokesperson for Private Fuel Storage LLC, Telephone Interview, 22 March 2002.

(Martin, 2002b) Martin, Sue. Spokesperson for Private Fuel Storage LLC, Telephone Interview, 5 September 2002.

(Martin, 2003a) Martin, Sue. Spokesperson for Private Fuel Storage LLC, Telephone Interview, 6 February 2003.

(Martin, 2003b) Martin, Sue. Spokesperson for Private Fuel Storage LLC, Telephone Interview, 8 October 2003.

(Martin, 2004) Martin, Sue. Spokesperson for Private Fuel Storage LLC, Telephone Interview, 23 January 2004.

(Moore, 2000) Moore, Robert. "Public Service, Personal Gain in Wyoming," Center for Public Integrity, 21 May 2000 [http://www.50statesonline.org/cgi-bin/50states/states.asp?State=WY&Display=StateSummary]

(MITYC, 2005) *Joint Convention on the Safety of Spent Fuel Management and on the Safety of Radioactive Waste Management*, Second Spanish National Report, October 2005 [http://www.mityc.es/NR/rdonlyres/BF3E47F5-7861-4A28-8FB3-8F2DBAB183A3/0/ConvencionInforme2_ing.pdf]

(MSDS, 2005) *Sweden's Second National Report Under the Joint Convention on the Safety of Spent Fuel Management and on the Safety of Radioactive Waste Management*, Ministry of Sustainable Development, Ds 2005:44 [http://www.sweden.gov.se/content/1/c6/05/40/89/fc570cf2.pdf]

(NEI, 1998) *Industry Spent Fuel Storage Handbook*, Nuclear Energy Institute, NEI 98-01, May 1998.

(NEI, 2002) "Appellate Court Ruling Likely Makes Taxpayers Liable for DOE Failure to Dispose of Used Nuclear Fuel," News, Nuclear Energy Institute, 2002 [http://www.nei.org/doc.asp?Print=true&DocID=980&CatNum=&CatID=]

(NO Coalition, 2000) "White Paper Regarding Opposition to the High-Level Nuclear Waste Storage Facility Proposed By Private Fuel Storage On The Skull Valley Band of Goshute Indian Reservation, Skull Valley, Utah," The Coalition Opposed to High-Level Nuclear Waste, 28 November 2000.

(NRC, 1996) "Strategic Assessment Issue: 6. High-level Waste and Spent Fuel," Nuclear Regulatory Commission, 16 September 1996 [http://www.nrc.gov/NRC/STRATEGY/ISSUES/dsi06isp.htm]

(NRC, 2001) "Final Environmental Impact Statement for the Construction and Operation of an Independent Spent Fuel Storage Installation on the Reservation of the Skull Valley Band of Goshute Indians and the Related Transportation Facility in Tooele County, Utah," U.S. Nuclear Regulatory Commission, NUREG-1714, vol. 1, December 2001: 7-1 thru 7-5.

(NRC, 2002) "NRC Publishes Final Environmental Impact Statement on Proposed Spent Fuel Storage Facility," *NRC News*, Nuclear Regulatory Commission, No.02-002 [http://www.nrc.gov/reading-rm/doc-collections/news/archive/02-002.html]

(NRC, 2004) "NRC Licenses Spent Nuclear Fuel Storage Facility at Idaho National Engineering and Environmental," *NRC News*, U.S. Nuclear Regulatory Commission, No. 04-150, 1 December 2004 [http://www.nrc.gov/reading-rm/doc-collections/news/2004/04-150.htm].

(NRC, 2005a) "Memorandum and Order," Nuclear Regulatory Commission, Docket No. 72-22-ISFSI, CLI-05-19, 9 September 2005 [http://www.nrc.gov/reading-rm/doc-collections/commission/orders/2005/2005-19cli.pdf]

(NRC, 2005b) "NRC Commissioner Jaczko Takes Oath of Office; Commissioner Lyons Swearing-In Set for Next Week," *NRC News*, U.S. Nuclear Regulatory Commission, No. 05-013, 21 January 2005 [http://www.nrc.gov/reading-rm/doc-collections/news/2005/05-013.pdf]

(NRC, 2005c) "NRC Denies Utah's Final Appeals, Authorizes Staff to Issue License for PFS Facility," *NRC News*, U.S. Nuclear Regulatory Commission, No. 05-126, September, 2005 [http://www.nrc.gov/reading-rm/doc-collections/news/2005/05-126.html]

(Nuke-Energy, 2002) "Federal Judge Rules Against Utah Nuclear Waste Laws," [http://www.nuke-energy.com/data/stories/aug02/fed%20judge.htm]

(NWPA, 1982) Nuclear Waste Policy Act of 1982 (P.L. 97-425) [http://www.rw.doe.gov/progdocs/nwpa/nwpa.htm]

(NWTRB, 1996) *Disposal and Storage of Spent Nuclear Fuel – Finding the Right Balance*, Nuclear Waste Technical Review Board, March 1996 [http://www.nwtrb.gov/reports/storage.pdf]

(OCEP, 2000) "Owl Creek Energy Project: Private Interim Spent Fuel Storage Facility," Owl Creek Energy Project, Global Spent Fuel Management Summit, 3-6 December 2000.

OCRWM (2001a) "Spain's Radioactive Waste Management Program," Office of Civilian Radioactive Waste Management, U.S. Department of Energy, DOE/YMP-0415, June 2001 [http://www.ocrwm.doe.gov/factsheets/doeymp0415.shtml]

OCRWM (2001b) "Sweden's Radioactive Waste Management Program," Office of Civilian Radioactive Waste Management, U.S. Department of Energy, DOE/YMP-0416, June 2001 [http://www.ocrwm.doe.gov/factsheets/doeymp0416.shtml]

OCRWM (2001c) "Switzerland's Radioactive Waste Management Program," Office of Civilian Radioactive Waste Management, U.S. Department of Energy, DOE/YMP-0417, June 2001 [http://www.ocrwm.doe.gov/factsheets/doeymp0417.shtml]

(PFS, 2000) "The PFS Facility," Private Fuel Storage, LLC [http://www.privatefuelstorage.com/project/facility.html]

(PFS, 2001a) "Motions Filed in Skull Valley Band/Private Fuel Storage vs. State of Utah," News Release 12 December 2001, Private Fuel Storage, LLC [http://www.privatefuelstorage.com/whatsnew/newsreleases/nr12-12-01.html]

(PFS, 2001b) "The Licensing Process," Private Fuel Storage, LLC [http://www.privatefuelstorage.com/project/licensing.html]

(PFS, 2003a) "Private Fuel Storage Challenges Licensing Board Decision," Private Fuel Storage, LLC, News Release 31 March 2003 [http://privatefuelstorage.com/whatsnew/newsreleases/nr3-31-03.html]

(PFS, 2003b) "Licensing Board Denies on a Technicality PFS's 'Smaller Site' Proposal," Private Fuel Storage, LLC, [http://www.privatefuelstorage.com/whatsnew/whatsnew.html]

(PFS, 2005a) "Atomic Safety and Licensing Board Recommends License for Spent Nuclear Site: First Such Recommendation in Nearly a Decade," News Release 24 February 2005, Private Fuel Storage, LLC [http://www.privatefuelstorage.com/whatsnew/newsreleases/nr2-24-05.html]

(PFS, 2005b) "Private Fuel Storage Wins Approval from Nuclear Regulatory Commission," News Release 9 September 2005, Private Fuel Storage [http://www.privatefuelstorage.com/whatsnew/newsreleases/nr9-09-05.html]

(*PFS v. Utah*, 2002) Case No. 2:01-CV-270C, 30 July 2002 [http://web.lexis-nexis. com/universe/printdoc]

(Public Citizen, 2001) "Another Nuclear Rip-off: Unmasking Private Fuel Storage," Public Citizen's Critical Mass Energy & Environment Program, July 2001.

(Roche, 2002) Roche, Lisa Riley. "Judge won't rule on wisdom of N-waste facility," *Deseret News*, 12 April 2002 [http://www.energy-net.org/IS/EN/NUKE/WST/ FED/NEWS/NP-1212.402]

(Ryan, 2003) Ryan, Cy. "Nuclear dump in Utah set for approval," *Las Vegas SUN*, 10 February 2003 [http://www.lasvegassun.com/sunbin/stories/text/2003/feb/10/ 514642537.html]

(Spangler and Davidson, 2002) Spangler, Jerry D. and Lee Davidson. "Utilities' promise to Bennett, Hatch may carry little weight," *Deseret News*, 14 July 2002 [http://deseretnews.com/dn/view/0,1249,405017719,00.html]

(Struglinski, 2005) Struglinski, Suzanne. "Nuclear transport to Utah may face problems," *Las Vegas Sun*, 12 September 2005 [http://www.lasvegassun.com/ sunbin/stories/sun/2005/sep/12/519342580.html?struglinski]

(Struglinski, 2006a) Struglinski, Suzanne. "Nuclear waste site looks doomed," *deseretnews.com*, 8 September 2006 [http:deseretnews.com/dn/print/ 1,1442,645199773,00.html]

(Struglinski, 2006b) Struglinski, Suzanne. "PFS is still optimistic: Firm's chief aims to get interim storage in Utah," *deseretnews.com*, 28 September 2006 [http: deseretnews.com/dn/print/0,1249,650194421,00.html]

(Stuart, 1999) Stuart, Ivan F. "Owl Creek Energy Project: A Solution to the Spent Fuel Temporary Storage Issue," Waste Management Conference, 28 February-4 March 1999.

(Talhelm, 2005) Talhelm, Jennifer. "Skull Valley waste site blocked, at least for now," *Daily Herald*, 16 September 2005 [http://www.harktheherald.com/modules. php? op=modload&name=News&file=article&sid=64418]

(trib.com, 2001) "Company still pushing nuclear dump," *Casper Star-Tribune*, Tuesday, 28 August 2001 [http://www.state.nv.us/nucwaste/news2001/nn11383. htm]

(U.S. Congress, 1997) H.R. 1270, House, 105th Congress, 1st Session, Bill Summary and Status, Vote 557, 30 October 1997 [http://thomas.loc.gov/cgi-bin/bdquery/ z?d105:HR01270:@@@X]

(U.S. Congress, 1998) H.R. 1270, Senate, 105th Congress, 2nd Session, Vote 148, 2 June 1998 [http://www.senate.gov/legislative/vote1052/vote_00148.html]

(U.S. Congress, 2000) S. 1287, Senate, 106th Congress, 2nd Session, Vote 88, 2 May 2000 [http://thomas.loc.gov/cgi-bin/bdquery/z?d106:SN01287:@@@D]

(Utah DEQ, 1997) "State of Utah's Contentions on the Construction and Operating License Application by Private Fuel Storage, LLC for an Independent Spent Fuel Storage Facility," Nuclear Regulatory Commission, Docket No. 72-22-ISFSI, 23 November 1997. Available at: [http://www.deq.utah.gov/Issues/no_high_level_ waste/documents/11_23_97.PDF].

(Utah DEQ, 2000) "Comments Submitted by the State of Utah on the Draft Environmental Impact Statement (DEIS)," Utah Department of Environmental

Quality, DEIS Comments Index, 20 September 2000 [http://www.deq.utah.gov/ Issues/no_high_level_waste/commentsindex.htm]

(Utah DEQ, 2002) "Information on Private Fuel Storage's Proposal to Locate a High Level Nuclear Waste Storage Facility on the Skull Valley Goshute Indian Reservation," State Concerns Statement, Utah Department of Environmental Quality [http://www.deq.utah.gov/Issues/no_high_level_waste/concerns.htm].

(*Utah v. PFS*, 2000) U.S. Court of Appeals for the Tenth Circuit, No. 99-4104, 25 April 2000. [http://web.lexis-nexis.com/]

(Utah.gov, 1997) "Governor announces opposition to proposed high level nuclear waste dump in Utah," 14 April 1997 [http://yeehaw.state.ut.us/]

(Utah.gov, 2001) "Governor Signs Legislation Blocking High-Level Nuclear Waste From Coming to Utah," 13 March 2001 [http://yeehaw.state.ut.us/]

Chapter 4

Permanent Disposal

The search for a long-term solution of what to do with the growing stockpiles of spent nuclear fuel from commercial nuclear utilities and high-level waste from defense facilities has been especially difficult. Paradoxically, as the need for action has become more acute, the conflict generated by the politicization of the issue has delayed progress on finding an answer to the problem.

This chapter discusses the search for a permanent solution for managing highly radioactive waste, briefly overviews how other countries are pursuing the long-term management of nuclear waste, examines the U.S. Yucca Mountain project in detail, and presents the advantages and the disadvantages associated with a permanent repository.

The Search for a Permanent Solution

There are two main characteristics of highly radioactive waste that must be overcome to ensure safe disposal: heat generation and high radioactivity. Heat generation limits the amount of nuclear waste that can be disposed in a given area, and the high radiation levels require shielding and remote handling systems. (IAEA, 2006b: 59) Thus, any disposal option must cope with these challenges. Over the years, a variety of proposals have been made on how best to dispose of highly radioactive waste: ocean/sub seabed, extraterrestrial, remote island, ice sheet, reprocessing/ transmutation, deep boreholes, and geologic repository. While there may be some advantages associated with each of these disposal options, each also has a number of disadvantages. Only the geologic repository option is being seriously considered today.

Three *ocean/sub seabed disposal* options have been proposed for highly radioactive waste. One proposal advocated dropping canisters filled with nuclear waste into the deepest parts of the ocean such as the Marianas Trench. A second option proposed dropping missile-shaped steel waste canisters from the surface that would penetrate up to 70 meters (approximately 229 feet) deep into the seabed sediment. A third proposal advocated dropping the waste into deep pre-drilled boreholes in the ocean floor. (NAS, 2001: 123) The advantage of ocean dumping is that the oceans comprise over 70 percent of the earth's surface, and the area is largely uninhabited. Thus, any release of radiation would be diluted substantially. However, there are a number of problems with ocean and sub-seabed disposal. First, current international law prohibits the dumping of radioactive waste at sea. The 1993 amendment to the *London Convention on the Prevention of Marine Pollution by Dumping of Wastes and Other Matter*, and the subsequent 1996 Protocol, prohibit the dumping of radioactive waste into the ocean. (London Convention, 1993) Second, changes in the seabed from the 2005 earthquake off the coast of Sumatra, Indonesia, and subsequent tsunami in the Indian Ocean, show why ocean/sub seabed disposal has substantial risks. The seabed was altered significantly in some places by the 9.0 plus quake, causing large thrust ridges (some almost a mile high) in some locations and diverted canyons in other places of the seabed. (Highfield, 2005: 1) Third, sea water is very corrosive and likely would significantly degrade any storage canister, which inevitably would cause the radioactive waste to leak out. It would be difficult to monitor and virtually impossible to repair any leaks. Fourth, this method of disposal would require the transport of significant quantities of highly radioactive waste across great distances to the remote nuclear waste dump. Finally, it would be difficult to retrieve the waste if a more effective method of disposal was found.

Extraterrestrial disposal also has been proposed as a means to permanently eliminate radioactive waste. (OCRWM, 2005: 1; NAS, 2001: 122) Basically, the highly radioactive waste would be launched by rocket directly into the sun. The advantage of space disposal is that it would be permanent and not require monitoring. However, the risk of an accident during launch and the subsequent release of radiation into the atmosphere are unacceptably high. Moreover, the use of such a disposal method would not be cost effective.

Remote uninhabited island disposal (including artificial islands) is another option that has been proposed. (OCRWM, 2005: 1; NRC, 2001: 124) The advantage of this method of disposal is that the waste would be far away from inhabited areas and thus, any radioactive release would have a minimal impact. However, there would be some risks. First, this method of disposal would require the transport of significant quantities of highly radioactive waste across great distances to the remote disposal repository. Second, there is the potential for a strong storm such as a typhoon or hurricane to cause damage to the disposal facility, which could result in a release of radioactive material. Third, since the radioactive waste would be disposed of in a remote location, it would be more difficult to monitor the waste and protect the material from theft. Finally, if global warming does result in a significant rise of the oceans, there would be the threat of submerging the disposal facility under water.

A fourth option is *ice sheet disposal* in the Antarctic or Arctic. (OCRWM, 2005: 1) The waste would be emplaced in shallow boreholes where heat from the decomposing nuclear material would cause the waste canisters to sink slowly into the ice. The advantage of this method of disposal is that it would be located in a remote location. However, there would be some risks too. First, this method of disposal would require the transport of significant quantities of highly radioactive waste across great distances to the remote nuclear waste dump. Second, global warming has caused the ice sheets to become unstable in some areas and have resulted in large quantities of ice to either melt or break off into the ocean. Thus, the waste containers eventually could become exposed. Third, since the radioactive waste would be disposed of in a remote location, it would be more difficult to monitor the waste and protect the material from theft. Finally, the harsh climate and remoteness of the facility would make the cost of this disposal method expensive.

Reprocessing (partitioning) and transmutation would not be radioactive waste disposal options per se, but both can reduce the inventories of spent nuclear fuel and the amount of time that the nuclear waste would remain a threat to health and the environment. (OCRWM, 2005: 1; NAS, 2001: 119-22) However, it would not alleviate the need to dispose of the radioactive waste eventually. *Reprocessing (or partitioning)* uses a chemical process to separate out the plutonium and fissionable uranium from the spent nuclear fuel rods. The advantage of this method is that it reduces the volume of highly radioactive waste. However, there are *proliferation* (the spread of nuclear material to an unauthorized third party) concerns associated with this method since it would produce plutonium and fissionable uranium. *Transmutation* involves reprocessing the spent nuclear fuel by altering the radioactivity of the elements so that they remain a radioactive threat for a shorter period of time. (OCRWM, 2005: 1) The advantage to this method is that the lethality of the waste is reduced. While both the reprocessing and transmutation options are potentially attractive, it is estimated that it would take over 100 years to reprocess the commercial spent fuel. (NAS, 2001: 121) Moreover, the scientific consensus is that while reprocessing and transmutation would reduce the volume of highly radioactive waste, it would not eliminate the need for the long-term management of highly radioactive waste. (NAS, 2001: 119).

The *deep borehole disposal* method would emplace the canisters containing the radioactive waste deep underground. This method is similar to geologic disposal but has the advantage of emplacing the waste in extremely deep levels well below any depth that a mined repository could achieve. The main disadvantage of this method is that the waste could not be retrieved easily if a more effective method of disposal was found. (OCRWM, 2005: 1)

Finally, geologic disposal is the most discussed option by engineers for disposing of highly radioactive waste. It appears that nature created the world's first geologic radioactive waste repository in West Africa over a billion years ago. Fifteen natural nuclear fission reactors have been discovered in Oklo, Gabon. Scientific tests indicate that the resultant plutonium from the natural chain reaction has moved less than 10 feet over the millennia; bolstering claims that a man-made geologic repository for highly radioactive waste could be capable of effectively isolating the lethal nuclear material for many years. (OCRWM, 2004c: 1, 3) *Geologic disposal* consists of isolating the highly radioactive waste by emplacing the waste canisters in a mined underground repository. A number of different types of geologic formations (e.g., clay, granite, salt) have been studied over the years to determine what would be the best medium to dispose of the waste. According to the National Academy of Sciences (NAS), geologic disposal is the only scientifically and technically credible option that can provide long-term safety without active management. (NAS, 2001: 27) The advantage of this method is that the waste would be secured for the long-term. Moreover, the repository could be constructed in a manner that allows retrieval of the waste in the future. However, there are societal and political implications associated with this method of disposal.

National Approaches to Highly Radioactive Waste Disposal

Today, thirteen countries are actively involved in pursuing long-term management programs for radioactive waste. However, no country has completed a geologic disposal facility. Of these countries, the United States likely will operate the world's first geologic repository for highly radioactive waste at Yucca Mountain, Nevada. Although, selection of the repository site has been highly controversial which, in turn, has delayed completion of the project until 2017-2020 at the earliest. Sweden possibly could complete its highly radioactive waste disposal facility before then. Finland is projected to follow shortly thereafter. The remaining countries are not scheduled to build repositories until sometime in the distant future (see Table 4.1 – Countries with Active Highly Radioactive Waste Management Programs).

Table 4.1 Countries with Active Highly Radioactive Waste Management Programs

Country	Projected Repository Opening
Belgium	after 2035
Canada	after 2034
China	at earliest 2040
Finland	projected 2020
France	after 2020
Germany	no projected date
Japan	at earliest 2035
Russia	to be determined
Spain	to be determined
Sweden	around 2015
Switzerland	after 2050
United Kingdom	to be determined
United States	at earliest 2017

Source: Adapted from OCRWM (2001k) "Radioactive waste: an international concern," Office of Civilian Radioactive Waste Management, U.S. Department of Energy, DOE/YMP-0405, June 2001[http://www.ocrwm.doe.gov/factsheets/doeymp0405.shtml]

It should be noted that not all countries believe that disposal is the preferred method of coping with the mounting inventories of radioactive waste. Some countries consider spent nuclear fuel as a resource that should be reprocessed. While reprocessing will reduce highly radioactive waste inventories, it is important to note that it will not preclude the need for long-term management such as disposal since it will not totally eliminate the lethal waste.

For those countries that do follow a long-term highly radioactive waste disposal policy there generally are several common features to their disposal programs. First, decision makers have been slow to address the problem of the mounting inventories of highly radioactive waste generally due to political opposition. Second, the "polluter pays" principle is followed (i.e., the nuclear waste generator is responsible for paying for disposal costs). Special disposal funds financed by the nuclear utilities have been set up to pay for feasibility studies and operating costs. Third, independent agencies have been established to be responsible for the final disposal of nuclear waste. Fourth, the multi-barrier approach is being followed – using both natural and man-made engineered barriers. Finally, a deep underground geologic repository is the preferred method for disposal.

Belgium has generated almost 3,000 metric tons of spent nuclear fuel and is expected to produce an additional 1,500 metric tons of spent fuel until the last commercial nuclear power facility is closed. (EIA 2001:1; FANC 2006: 9) Disposal in a deep geologic repository presently is the preferred method of long-term management for highly radioactive waste. (FANC 2006:13-14) A sedimentary clay area in north-east Belgium (Boom Clay) has been identified as a possible repository site. The current focus is to use a multi-barrier design approach using the clay formation as a natural barrier and adding a man-made engineered barrier system to enhance confidence in the feasibility of the repository to contain the radioactivity. (FANC 2006:15) A national repository for highly radioactive waste is not expected to be ready until after 2035 – possibly as late as 2080. (OCRWM 2001a:1-2)

Canada has produced approximately 34,000 metric tons of spent nuclear fuel and is expected to add an additional 25,000 metric tons by 2020. (EIA 2001:1) Canada refers to spent nuclear fuel as "spent fuel waste." Currently, the spent fuel waste is held in interim storage pending a decision on which long-term management method to implement. (CNR 2005:9) Two main principles govern Canadian radioactive waste disposal: (1) use of multiple engineered barriers to ensure containment and isolation, and (2) procedures to monitor the performance of the engineered barriers. (CNR 2005:53) Canada explored the viability of a deep underground repository (the Underground Research Laboratory) for containing radioactive waste within a granite rock mass. (URL 2006:1) A 1998 report concluded that a deep geologic disposal of highly radioactive waste 500 to 1000 meters deep within stable rock would provide a safe long-term storage option for highly radioactive waste. In spite of this progress, however, Canada is not expected to have an operational permanent highly radioactive waste disposal facility until sometime after 2034. (OCRWM 2001b:3)

China is expected to be the fifth largest generator of nuclear power behind the USA, France, Japan and South Korea. While China only produced about 730 metric tons of spent nuclear fuel by 2005, future projections predict almost a ten-fold increase to over 6,000 metric tons by 2020 as China aggressively plans to expand its nuclear power generation. (EIA 2001:1) Thus, as inventories of highly radioactive waste continue to accumulate, there will be increased pressure on Chinese policymakers to address the mounting inventories of highly radioactive waste. Because China only recently began to build a civilian nuclear power program (the first electricity generating reactor came on-line in 1991), radioactive waste disposal has not been a top priority for the government. Instead, efforts have been directed toward aggressively expanding energy production to keep pace with the rapidly growing economy. China is not expected to have an operational permanent highly radioactive waste disposal facility until 2040 at the earliest. (OCRWM 2001c:1)

Almost 1,800 metric tons of spent fuel has been produced in Finland, and an additional 1,000 metric tons are expected to be produced by 2020. (EIA 2001:1) In 2004, site characterization studies on the feasibility of the disposal facility were begun at the ONKALO underground rock characterization facility at the Olkiluoto site. The multi-barrier repository design will consist of a network of tunnels constructed at a depth of 400-500 meters. The cast iron and copper disposal canisters will be surrounded by a bentonite backfill to act as a sealant. (STUK 2005:17) If found suitable, the repository is expected to be ready to start accepting up to 5,700 metric tons of nuclear waste around 2020. (STUK 2005:16; OCRWM 2001d:3)

France has generated approximately 36,000 metric tons of spent nuclear fuel. Inventories are expected to increase to about 51,000 metric tons by 2020 (behind only the USA, Canada and the UK). (EIA 2001:1) France does not consider spent fuel as waste and actively engages in reprocessing. Thus, France's total waste inventories are lower than would be expected since it reprocesses spent fuel rods which reduce the amount of highly radioactive waste generated. In 1998, France began a dual track process in its search for a permanent repository: move forward on a clay site and begin a search for a new site with outcropping granite. (NEA 2006:18) Disposal would take place in a repository. The repository concept is to be based on principle of reversibility – the ability to reverse the process at any time and retrieve the highly radioactive waste. However, France is not expected to have an operational permanent highly radioactive waste disposal facility until sometime after 2020. (OCRWM 2001e:3)

Germany has produced almost 12,000 metric tons of spent nuclear fuel and is expected to produce an additional 5,500 metric tons of highly radioactive waste by 2020 for a total inventory of 17,500 metric tons. (EIA 2001:1) As the environmental movement strengthened in Germany with the advent of the rise of the Green Party, opposition to nuclear power began to increase. In 2000, Germany decided to abandon nuclear generated power altogether. By 2005, over half of the nuclear power plants in Germany have either been dismantled or were in the process of being dismantled. (BMU 2005:9) While there is no set date for completion of an operational permanent highly radioactive waste disposal facility in Germany, the federal government hopes to establish a deep geologic repository by 2030. (BMU 2005:14) However, the political realities today are such that it is unlikely that a repository will be ready to accept nuclear waste by then.

Japan has generated about 23,000 metric tons of spent nuclear fuel, and is expected to produce an additional 21,000 metric tons of highly radioactive waste by 2020. (EIA 2001:1) Japan's official policy to cope with their increasing inventories of highly radioactive waste is to reprocess spent fuel. (IAEA 2006a:9) While this reduces the amount of highly radioactive waste, it does not totally eliminate it since highly radioactive waste is a by product of the reprocessing cycle. Currently, research into the feasibility of a repository design is being conducted at the Horonobe Underground Research Laboratory Project on the northern island of Hokkaido. Japan considers deep geologic disposal as the only suitable method for the long-term management of highly radioactive waste. (METI 2005:B-3) However, Japan is not expected to have an operational disposal facility ready until after 2035 at the earliest. (OCRWM 2001f:3)

Russia has produced over 22,000 metric tons of spent nuclear fuel and is expected to produce an additional 10,000 metric tons of highly radioactive waste by 2020. (FAEA 2006:53; EIA 2001:1) Russia does not view spent nuclear fuel as radioactive waste, but instead as a resource to reuse. Spent fuel is either to be reprocessed or stored. Nevertheless, Russia has been investigating the feasibility of salt, granite, clay and basalt as possible host geologic mediums for a possible future repository. However, at present, there is no projected date for completion of an operational permanent highly radioactive waste disposal facility in Russia. (OCRWM 2001g: 3)

Spain has generated over 5,000 metric tons of spent nuclear fuel and is expected to produce an additional 2,000 metric tons of highly radioactive waste by 2020. (EIA 2001:1) Although Spain initially intended to reprocess spent nuclear fuel, the government ultimately decided not to reprocess nuclear waste. However, a decision on how to pursue long-term highly radioactive waste management has been postponed until after 2010. Instead, Spain is focusing on centralized interim storage. Thus, there is no projected date for completion of an operational permanent highly radioactive waste disposal facility in Spain. (MITYC 2005:6; OCRWM 2001h:3)

Sweden has produced about 5,000 metric tons of spent nuclear fuel and is expected to produce an additional 2,000 metric tons by 2020. (EIA 2001:1) After approximately a year of being in a spent fuel pool, spent nuclear fuel is transported to a centralized interim storage facility from reactor spent fuel pools and will be stored for about 30 years prior to being transported to a permanent repository. Currently, an active search is being conducted for a permanent repository site. While no specific site has yet been identified, the repository for highly radioactive waste is envisioned to be a multi-barrier repository design. Sweden is not expected to have an operational permanent highly radioactive waste disposal facility until after 2015, probably around 2020. (IAEA 2006b:63)

Switzerland has generated almost 2,000 metric tons of spent nuclear fuel and is expected to produce an additional 1,000 metric tons of highly radioactive waste by 2020. (EIA 2001:1) Switzerland views spent nuclear fuel as a resource that can be reprocessed and as of 2005 no spent fuel has been declared as waste. (HSK 2005:49) Initially, Switzerland studied the feasibility of crystalline rock for a potential repository. However, that site was not found suitable; now the Opalinus clay formation in the northern part of the country is being considered. (HSK 2005:9) However, according to the Swiss government, a repository will not be needed for "several decades." (HSK 2005:8) Since no repository has been constructed, highly radioactive waste not stored on-site is transported to a centralized interim storage facility. An operational permanent highly radioactive waste disposal facility is not expected to be ready until after 2050. (OCRWM 2001i:3)

The UK has produced over 47,000 metric tons of spent nuclear fuel and is expected to produce an additional 11,000 metric tons of highly radioactive waste by 2020. (EIA 2001:1) The UK reprocesses its own spent nuclear fuel which has reduced the overall amount of highly radioactive waste. No spent fuel has been designated as radioactive waste for disposal. (Defra 2006:65) In July 2006, a government sponsored report recommended geologic disposal for the highly radioactive waste that has been generated. There is no projected date for completion of an operational permanent highly radioactive waste disposal facility in the United Kingdom. (OCRWM 2001j:3)

Since no highly radioactive waste disposal facility is expected to be ready until 2017 at the earliest, it is apparent that the world will be awash in nuclear waste. Only the USA and Finland have selected a repository site; Sweden has narrowed its search to two sites. (IAEA 2006b:61) Even after deep geologic repositories come on-line in the USA, Finland and Sweden, it will be difficult to reduce highly radioactive waste inventories since the amount generated annually will account for much of the waste being transferred to a disposal facility.

History of Establishing a Geologic Repository in the U.S.

It has been exceedingly difficult for U.S. policymakers to develop a broad consensus on how to approach the radioactive waste disposal issue (see Table 4.2 – History of Highly Radioactive Waste Disposal). Initially, it was difficult to get someone even to address the topic. When it finally became apparent that the storage and disposal problem could no longer be ignored, the process to develop a long-term solution for radioactive waste has been slow and arduous. The policy process has been quite contentious since the costs are immediate and tangible, while the benefits are long-term and difficult to measure. In addition, the inherent tension between a state's ability to exercise sovereignty within its own borders and the federal government's actions to construct and operate a repository for highly radioactive waste further complicate the issue. Moreover, since highly radioactive waste disposal has not been a salient issue with the public, there has been little incentive for policy makers to address the controversial issue. Thus, by default, the policy of simply storing the highly radioactive waste on-site at the point of origin evolved (either in cooling pools or above ground in dry cask storage). However, this is not a viable long-term solution.

The Early Years

Initially, in the 1940s and early 1950s, only small amounts of radioactive waste were generated, principally as a byproduct of the U.S. atomic and nuclear weapons programs. Even after the development of commercial nuclear power production, little attention was paid to how to dispose of the waste safely; during this period, commercial nuclear reactors and the defense industry did not produce enough radioactive waste to cause great concern. There was little discussion about storage or disposal options, and no provisions for permanent disposal were established. At that time, policymakers did not consider radioactive waste to be an issue that needed to be addressed.

Table 4.2 History of Highly Radioactive Waste Disposal

1957–National Academy of Sciences recommends disposing of highly radioactive waste underground in salt mines or salt domes.

1960s–Focus shifts from storage/disposal to the reprocessing of spent nuclear fuel.

1966–First commercial reprocessing plant opens in New York (plant forced to close down early 1970s) due to technical problems. Other similar facilities in South Carolina and Illinois also closed.

1970–Atomic Energy Commission proposes building a permanent repository in salt deposits near Lyon, KS. Proposal eventually abandoned due to environmental concerns.

1975–Ford Administration decides to forego reprocessing due to proliferation concerns.

1979–Interagency group recommends to Carter Administration to proceed with geologic disposal.

1981–Reagan Administration initially reverses reprocessing policy, but later also abandons reprocessing due to technical problems, cost, and proliferation concerns.

1982–Nuclear Waste Policy Act (NWPA) mandates a process of establishing a national repository.

1983–Nine sites in six western states (LA, MS, NV, TX, UT and WA) identified as a potential candidate for a repository.

1985–President Reagan decides that a defense-only repository is not necessary and that defense-generated highly radioactive waste could be disposed along with spent nuclear fuel from civilian reactors.

1986–Of original nine sites, five identified as a possible candidate for a national repository (Richton Dome, MS; Yucca Mountain, NV; Deaf Smith County, TX; Davis Canyon, UT; and, Hanford, WA). Three eventually selected for detailed examination (Yucca Mountain, Deaf Smith County, and Hanford). Search for a site in the East abandoned due to intense political pressure from eastern states.

1987–NWPA amended which singles out Yucca Mountain as the only site to be studied.

1992–Energy Policy Act requires EPA to establish health and safety standards for Yucca Mountain.

1996–Federal appeals court rules that federal government must comply with 1982 NWPA timeframe, but did not specify what would happen if deadline missed. Subsequent ruling allowed utilities to seek damages for non-compliance with deadline.

2002–Bush Administration recommends pursuing Yucca Mountain. Nevada issues a notice of disapproval. Congress overrides Nevada "veto."

2005–Administration announces that the 2010 date for opening the repository cannot be met. Correspondence from U.S. Geological Survey employees indicate fraudulent data on water flow rates may have been submitted.

2006–Yucca Mountain expected to receive radioactive waste in 2017 at the earliest.

In the mid-1950s, at the request of the Atomic Energy Commission, the National Academy of Sciences was asked to study the highly radioactive waste disposal issue. In 1957, an NAS panel recommended that the radioactive waste be stored permanently in deep underground salt beds. (NAS, 1957: 4) Four potentially suitable regions were identified:

- salt domes of the Gulf Coastal Plain in Texas, Louisiana and Mississippi
- bedded salt in the Paradox Basin of Colorado, Arizona and New Mexico
- bedded salt in the Permian Basin of Kansas, Oklahoma, Texas, and New Mexico
- bedded salt in the Michigan and Appalachian Basins of Michigan, Ohio, Pennsylvania, and New York. (OCRWM, 2000: 65)

However, since radioactive waste was not yet perceived to be a salient issue, no serious steps were taken to implement the recommendation.

During the 1960s, the focus shifted from storage/disposal to the reprocessing of spent nuclear fuel (this was expected to reduce the overall amount of radioactive waste). In 1966, the first commercial reprocessing plant opened in West Valley, New York. Because the commercial reactors were not generating much spent fuel during this era, the West Valley plant principally processed spent fuel from defense industry reactors. However, technical problems and worker exposure to radiation forced the plant to close in the early 1970s. While, an additional reprocessing facility was built near Barnwell, South Carolina, it also was eventually forced to close. (Holt, 1998: 5) Another reprocessing plant was constructed in Morris, Illinois. However, technical problems precluded opening the plant for operations. Interestingly, the Morris reprocessing plant became the nation's first de facto highly radioactive waste storage facility when the plant accepted spent fuel from the commercial nuclear utilities that had signed contracts to reprocess the spent fuel. The radioactive waste is still stored at the Morris facility today. (Holt, 1998: 4-5) In addition to the Morris reprocessing facility, the West Valley (New York) plant also continues to store a small amount of radioactive waste. The Department of Energy nuclear weapons sites at Hanford (Washington) and Idaho National Engineering Laboratory also are storing small amounts of spent nuclear fuel. (Macfarlane, 2001: 211)

The 1970s

In 1970, the Atomic Energy Commission (AEC) proposed building a repository in salt deposits near Lyon, Kansas. However, the Lyon site was abandoned two years later after it was discovered that numerous oil and gas exploration boreholes had been drilled throughout the area.

A shift in U.S. radioactive waste policy took place in the mid-1970s. Mounting concerns about the potential diversion of weapons grade radioactive material during the reprocessing cycle prompted the Ford and Carter Administrations to abandon reprocessing efforts. Bending to industry concerns that this decision would ultimately lead to increased accumulations of spent fuel at commercial reactor sites, DOE broadened its efforts to site a permanent repository for highly radioactive waste. However, states being considered as a potential host for a radioactive waste disposal site opposed federal inquires on site suitability and the issue of a permanent repository became increasingly controversial and politicized. In an effort to reduce industry concerns over the accumulating spent fuel, the White House proposed to take title to the radioactive waste and provide federal interim storage until a permanent repository was available. (Holt, 1998: 5) Eventually, the

Carter Administration modified this proposal and attempted to persuade the industry to develop privately run storage facilities at former reprocessing plants. Neither the commercial nuclear power industry nor the states that had former reprocessing plants supported the Carter plan. One of the principal reasons for the lack of support for the Administration's interim storage proposal was the lack of progress in finding a suitable solution for siting a permanent repository for radioactive waste. Both the commercial nuclear power industry and the states with reactors were concerned that their locations would become de facto permanent storage sites.

The 1980s

The Reagan Administration initially reversed the Carter decision to abandon the reprocessing of spent nuclear fuel. However, technical problems with the reprocessing cycle, the fact that it was cheaper to mine new uranium rather than reprocess spent fuel, and continued concerns over the proliferation of highly radioactive weapons grade material to third parties, eventually led the U.S. to discontinue the reprocessing option altogether. (Holt, 1998: 4–5) The U.S. was now forced to resume its search for an acceptable method for radioactive waste disposal.

In 1982, the Nuclear Waste Policy Act began the process of establishing a deep, underground permanent national repository for highly radioactive waste. The Act mandated that DOE would first identify a site in the western U.S. followed by the designation of an additional site east of the Mississippi river. The 1982 Act also directed DOE to establish criteria for the site selection process, study multiple sites in the West, and select a location to be ready for disposal by January 1998. States that were identified as potential hosts for the radioactive waste storage facility had a history of some type of prior involvement with nuclear technology. Initially, nine sites in six states (Louisiana, Mississippi, Nevada, Texas, Utah and Washington) were identified in 1983 by DOE as a potential repository for radioactive waste (see Figure 4.1–Repository Sites Considered).[1]

Eventually, the list was narrowed down to five sites in five states (Richton Dome, Mississippi; Yucca Mountain, Nevada; Deaf Smith County, Texas; Davis Canyon, Utah; and, Hanford, Washington) after environmental impact studies were completed. In 1986, the list of potential sites for a national repository was cut to three locations: Yucca Mountain, Deaf Smith County and Hanford. (Rogers and Rogers, 1998: 62) Undoubtedly, DOE had calculated that a prior nuclear involvement would tend to mute any opposition to siting a radioactive waste storage facility proposal within a

---

1   See, "Repository Sites Considered in the United States." Yucca Mountain Project, Office of Radioactive Waste Management, U.S. Department of Energy, for an interactive map that shows each of the sites that were considered: http://www.ocrwm.doe.gov/ym_repository/about_project/sitesconsidered.shtml. The original nine sites were: "Vacherie Dome, Louisiana (salt dome); Richton Dome, Mississippi (salt dome); Cyprus Creek Dome, Mississippi (salt dome); Deaf Smith County, Texas (bedded salt); Swisher County, Texas (bedded salt); Davis Canyon, Utah (bedded salt); Lavender Canyon, Utah (bedded salt); Yucca Mountain, Nevada (volcanic tuff); and Hanford, Washington (basalt)." (OCRWM, 2005: 1)

given state.[2] However, in spite of DOE's hopes, the site selection process was slow and contentious due to the substantial opposition generated by the western states identified as a potential host for a repository.

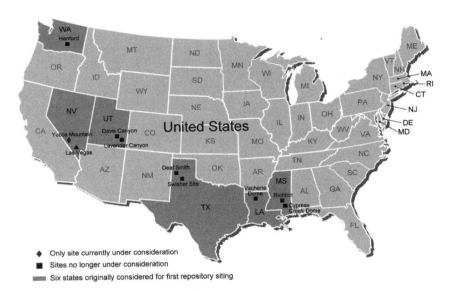

Figure 4.1 Repository Sites Considered
Source: OCRWM, Yucca Mountain Project, [http://www.ocrwm.doe.gov/ym_repository/about_project/sitesconsidered.shtml]

Although the vast majority of highly radioactive waste is generated east of the Mississippi, the site selection process in the East was even more difficult than in the West. Public opposition intensified after twelve sites in seven states (Georgia, Maine, Minnesota, New Hampshire, North Carolina, Virginia and Wisconsin) were selected to be studied for a repository east of the Mississippi River. (Kraft, 1996: 115) And, although the 1982 Act mandated that an eastern site would be designated at a later date, public and political opposition, as well as an approaching election forced the Reagan Administration to abandon the siting process in the East altogether in 1986. In fact, U.S. Representative Jim Matheson (R-UT) commenting on the fact that the only sites considered were in the West stated, "It tells you the politics are driving this more than the science perspective." (Abrahms, 2005: 3)

To circumvent further opposition, Congress passed the 1987 Nuclear Waste Policy Amendments Act (NWPAA) which singled-out Yucca Mountain, Nevada, as the only site to be studied for a proposed national repository for the country's highly

---

2   Yucca Mountain, Nevada, was adjacent to the Nevada Test Site; Deaf Smith County, Texas, hosted the Pantex Nuclear Weapons Assembly Plant; and Hanford, Washington, hosted a nuclear weapons production facility. (Kraft, 1996: 114-15).

radioactive waste.[3] The legislation that identified Yucca Mountain was primarily due to the result of efforts by congressional delegations from states that were being considered as potential host states for the national repository.[4] Clearly, the site selection of Yucca Mountain was motivated by political considerations. In fact, many observers have criticized the site selection process, and even some members of Congress openly admitted that the 1987 legislation was driven by political, not environmental, considerations. First, the eastern states had far more political clout in Congress than western states since they had larger congressional delegations. Thus, the burden for site selection shifted to only western states. Second, Nevada had a small population and more limited political influence than the other western finalists (Texas and Washington) which had larger and more influential congressional delegations.[5]

The 1990s

The 1990s witnessed little movement towards a permanent resolution of the nuclear waste issue. While Congress did pass the Energy Policy Act of 1992, which required the Environmental Protection Agency (EPA) to establish health and safety standards for the Yucca Mountain repository (e.g., the Act restricted the release of radioactive materials into the environment for at least 10,000 years), little else was accomplished. The Clinton Administration moved cautiously on the Yucca Mountain project due to environmental and technical concerns, as well as mounting political opposition. At the same time, Nevada worked feverishly to block the project at every juncture. In addition, a Government Accountability Office (GAO) study initiated at the request of Senator Harry Reid (D-NV) was highly critical of the site characterization process. The report also stated that a recommendation to proceed with the project "may be premature" since there were almost 300 outstanding technical issues.[6] (GAO, 2001: 3) The GAO believed that DOE would be unable to submit an acceptable license application to the Nuclear Regulatory Commission (NRC) within the time limits prescribed by the Nuclear Waste Policy Act. While the federal government

---

3   The 1982 Nuclear Waste Policy Act with the 1987 Amendments is available at: http://www.ocrwm.doe.gov/documents/nwpa/css/nwpa.htm.

4   Louisiana's influential Senator J. Bennett Johnston (D), who was chairman of the Senate Committee on Energy and Natural Resources and a member of both the Senate Appropriations and Budget Committees, was instrumental in inserting legislation that initially restricted the site selection process in the West to only one site which put Yucca Mountain at the top of the list. (Rissmiller, 1993: 106-7) Later, in the conference committee which was convened to reconcile the Senate and House versions of the legislation, Yucca Mountain was singled out as the only site to be studied.

5   At the time, Nevada had only two representatives, and its senators (Jacob Hecht–R, and Harry Reid–D) were both in their first term. In contrast, Texas had 27 representatives including Speaker of the House Jim Wright and influential senators (Lloyd Bentsen–D, and Phil Gramm–R). Washington's congressional delegation had eight representatives including influential Democrat Tom Foley who was the House Majority Whip (later Speaker of the House).

6   The GAO report is available at: [http://www.gao.gov/new.items/d02191.pdf].

continued investigating the suitability of Yucca Mountain for the disposal of highly radioactive waste, increased opposition from the State of Nevada slowed the process considerably. The Nevada congressional delegation had some success in reducing funding for Yucca Mountain. Clearly, Nevada officials reasoned that if they could delay the process, the political dynamics might change sufficiently enough to derail the project altogether.

The continued delay in establishing a national nuclear waste repository led to additional attempts to modify the 1982 Nuclear Waste Policy Act and the subsequent 1987 amendments. In 1996, the Senate attempted on two occasions (U.S. Congress, S. 1271 and U.S. Congress, S. 1936) to amend the Nuclear Waste Policy Acts. Both bills would have continued the process of establishing a permanent national repository at Yucca Mountain as well as designating an interim nuclear waste facility in Nevada. However, the House failed to pass similar legislation due to a threatened presidential veto. The Clinton Administration opposed establishing an interim facility until the feasibility of a permanent storage facility was determined. Congressional Democrats were reluctant to override any veto which would further weaken the president politically. In April 1997, the Senate passed the Nuclear Waste Policy Act of 1997 (U.S. Congress, S. 104). Initially, the Senate passed the legislation with a 65-34 majority, and the House passed similar legislation with a 307-120 majority. (U.S. Congress, 1997) However, the Clinton Administration announced that it would veto the 1997 Nuclear Waste Policy Act since it "would undermine the credibility of the Nation's nuclear waste disposal program by, in effect, designating a specified site for an interim storage facility before the viability of that site as a permanent geological repository has been assessed." (EOP, 1997: 1)   Since congressional Democrats were reluctant to further weaken the president's political power already compromised by a series of investigations by the independent special prosecutor, the proposed legislation was abandoned. One of the major provisions of this Act was to require DOE to build an interim storage facility at the Nevada Test Site (which is near Yucca Mountain). A similar bill was introduced in the House (U.S. Congress, H.R. 1270). Since this legislation was similar to the bills introduced in 1996, it also did not pass.

The delay in establishing a national repository for highly radioactive waste only has served to intensify conflict. In June 1994, 20 states and 14 utilities filed lawsuits against DOE to force the government to uphold its promise to accept nuclear waste by the original 1998 deadline (Holt and Davis 1996: 7). As former Energy Secretary Hazel R. O'Leary reportedly stated:

> I am mandated under the law to continue to explore Yucca Mountain to see if it should be the repository for our high-level nuclear waste... I have been sued by 24 states saying, "Look, lady, you've got a legal—and I say moral—commitment to take our spent nuclear fuel in 1998". (Nixon, 1995: 4)

A federal appeals court agreed. In July 1996, a three-judge panel ruled that the government must begin accepting radioactive waste from U.S. nuclear power plants in 1998 as originally mandated by the 1982 Nuclear Waste Policy Act even though a permanent storage facility would not be ready until well past the 1998 deadline. (Courier, 1996: A1) However, the Court ruling did not specify what would happen if DOE missed the 1998 deadline. Follow-up litigation by the nuclear utilities and states asked the courts to specify a remedy for DOE's noncompliance. Eventually, a U.S. Court of Appeals ruling stated that DOE would be liable for unspecified damages for the missed deadline, and a subsequent ruling would permit the commercial nuclear utilities to seek damages. (ENR, 1998: 1; Holt, 2001: 2-3) Suits filed by several utilities are still pending against DOE for several billion dollars in damages.[7] However, the utilities still have not been able to force DOE to take title to the waste. In 2000, the U.S. Supreme Court declined to hear an appeal by the commercial nuclear utilities which sought to overturn an earlier U.S. Court of Appeals ruling which did not force DOE to accept the spent nuclear fuel.

The New Millennium

The lack of real progress during the 1990s has impacted significantly on the project. In 2000, the Office of Civilian Radioactive Waste Management (OCRWM) acknowledged that they may not be able to meet the scheduled 2010 opening for Yucca Mountain due to budget cuts. (Nuke-Energy, 2002: 1) In 2001, the National Research Council recommended that the geologic disposal option be pursued with the understanding that the decision on whether or not to implement the disposal option be made after appropriate deliberation by the political leadership. (NAS, 2001: 19)

The arrival of the George W. Bush Administration prompted a shift in White House policy. In February 2002, Secretary of Energy Spencer Abraham recommended in a letter to the president approval of Yucca Mountain to the White House, citing "national interest" as a compelling reason to more forward with the project. Five areas were cited as relevant to national interests in Abraham's letter to the President. First, a permanent repository is needed to provide a "pathway" for radioactive waste from the Navy's nuclear powered fleet. Second, a repository would help promote non-proliferation objectives by disposing of weapons grade plutonium from decommissioned weapons. Third, national energy security (which relies on continued nuclear power production) is threatened by a lack of sufficient on-site radioactive waste storage. Fourth, the continued accumulation of radioactive waste (in over 100 sites in 39 states with over 160 million Americans living within 75 miles of the stored material) was cited as a concern for homeland security. Finally, Abraham cited the necessity to implement an environmentally sound plan for disposing of highly radioactive defense waste and spent nuclear fuel from commercial nuclear reactors. (Abraham, 2002: 2-3) Abraham's letter to the President also stated that the project was "scientifically and technically suitable for a repository." (Abraham, 2002: 1) The following day, President Bush notified Congress that the Administration intended to proceed with the Yucca Mountain project:

---

7   The GAO has noted that estimates of the cost of the delay in accepting the waste vary considerably from about $2 billion by DOE to $50 billion by the commercial nuclear utility industry. (GAO, 2001: 2)

Having received the Secretary's recommendation and the comprehensive statement of the basis of it, I consider the Yucca Mountain site qualified for application for a construction authorization for a repository. Therefore, I now recommend the Yucca Mountain site for this purpose. In accordance with section 114 of the Act, I am transmitting with this recommendation to the Congress a copy of the comprehensive statement of the basis of the Secretary's recommendation prepared pursuant to the Act. The transmission of this document triggers an expedited process described in the Act. I urge the Congress to undertake any necessary legislative action on this recommendation in an expedited and bipartisan fashion. (Bush, 2002: 1)

The Bush Administration's announcement to proceed with Yucca Mountain was predicated on two basic assumptions: national security and energy security. First, the tragic events of 9/11 and the concomitant terrorist threat against nuclear power plants has highlighted the vulnerability of the radioactive waste stored on-site to a potential terrorist attack. Second, the lack of resolution of the nuclear waste issue threatens the viability of the nuclear energy industry since some reactors could be forced to shut down due to a lack of adequate on-site storage for the spent nuclear fuel. Since the Bush national energy policy relies on the continued generation of nuclear power and promotes the construction of new nuclear power plants, any impediment to the continued production of commercial nuclear power would be perceived to be counterproductive to the Administration's energy plan.

In response to the White House decision to recommend Yucca Mountain, Governor Kenny Guinn (R-NV) exercised the right to oppose the president's decision by issuing the first "official notice of disapproval" of the Yucca Mountain site as outlined by the 1982 Nuclear Waste Policy Act. (Guinn, 2002: 1; LVRJ, 2002: 1) Referred to in the media as a "veto," the notice of disapproval requires both chambers of Congress to vote to override by a simple majority vote. Nevada intensely lobbied other members of Congress to oppose the designation of Yucca Mountain as a national repository for highly radioactive waste.

In spite of Nevada's efforts, in May 2002, the U.S. House of Representatives overwhelmingly supported President Bush's decision to move forward on Yucca Mountain by a 306-117 vote to override Nevada's "veto" of the president's decision. (U.S. Congress, 2002a) In July 2002, the Senate overrode Nevada's notice of disapproval by 60-39 allowing the Yucca Mountain project to proceed to the next phase. (U.S. Congress, 2002b) President Bush signed the legislation into law (P.L. 107-200) later that month. While Nevada valiantly tried to rally support for its opposition to the Yucca Mountain project, the obstacles in Congress were too great. Perceived national and energy security concerns, as well as support by members of Congress from states that either have nuclear reactors where the waste is piling up, or that could be identified as a potential host for a national repository for radioactive waste, provided the Administration with a comfortable majority.

Failing to convince the White House and Congress that the technical and environmental concerns were sufficient cause to abandon the Yucca Mountain project, Nevada has continued its fight in other ways. For example, in 2002, Nevada officials were able to convince the U.S. Conference of Mayors to pass a resolution opposing the transportation of highly radioactive waste across state borders until the federal government can guarantee the safety of the nuclear waste shipments.

(USCM, 2002: 1) While there clearly has been some concern displayed over the safety of the nuclear waste shipments – especially through communities with large population centers – it is difficult to measure the extent of the opposition by local officials since the issue is also being used as a vehicle to obtain additional funding from Congress.

While the Yucca Mountain project appeared to be back on track, several events in 2004 and 2005 again challenged the viability of the planned repository. In July 2004 a federal district court ruled that the EPA's 10,000-year regulatory compliance period was too short of a time-span to protect the public from radiation exposure. While opponents of the Yucca Mountain project lauded the court's ruling, supporters were dismayed since they believed that the 10,000 year period itself already was unrealistically too long. Supporters believe that the pace of technological innovation progresses so quickly that it is quite likely that a technological solution to the radioactive waste problem will be found in the not too distant future. In response to the court ruling, EPA issued new guidelines that would allow increased emissions after the 10,000-year period.[8] The increased dose rate allowed for the period after 10,000 years has drawn considerable criticism from opponents of the Yucca Mountain repository. In early 2005, two events took place that cast further doubt on the fate of the Yucca Mountain project. First, the OCRWM Director announced that the 2010 date for opening Yucca Mountain could not be met (Director Chu resigned shortly after the announcement). While no firm date for opening was reestablished at the time, some observers speculated that 2015 or 2020 might be more realistic. (LVRJ, 2005b: 1) Second, it was revealed by the Energy Department that U.S. Geological Survey (USGS) personnel may have fabricated water quality assurance data involving water infiltration and climate. (Bodman, 2005: 1) Two data sets reportedly were used and the actual test results supposedly were not submitted since they were outside the parameters outlined for the project. Instead, flow rate data that was within specified limits supposedly was supplied. (LVRJ, 2005a: 1-2) The charge of falsifying data seemed to put the viability of the Yucca Mountain project in jeopardy. Opponents of the project seized on the issue to call into question the geologic viability of Yucca Mountain as well as the trustworthiness of the federal government. In response, DOE undertook a study to reexamine previous work on water infiltration (i.e., water flow) rates. In 2006, DOE released results of the the the study that found that the water infiltration modeling work previously performed by USGS employees was not adversely affected by quality assurance problems. (DOE, 2006: 1)

In March 2007, citing national security and energy security concerns, the Bush Administration proposed the Nuclear Fuel Management and Disposal Act.[9] The Act would facilitate the licensing and construction of the Yucca Mountain repository

---

8   The allowable limit for the first 10,000 years is 15 millirem a year (slightly more than a chest X-ray), and then 350 millirem for up to 1 million years. (CBS News, 2005: 1)

9   See (Bodman, 2007) Letter to the Honorable Richard B. Cheney proposing the "Nuclear Fuel Management and Disposal Act," March 6, 2007 [http://www.ocrwm.doe.gov/info_library/newsroom/documents/2007bill. pdf] for a summary of the bill, and the proposed legislation.

by streamlining the NRC licensing process and reforming the budgetary process to provide adequate funding. The proposed Act also would eliminate the 70,000 metric ton limit for the repository in order to accommodate all of the country's highly radioactive waste, and would set aside a large tract of land in order to isolate the repository from the public. (DOE, 2007: 1)

The next step in the Yucca Mountain project is for DOE to submit a license application to the NRC. According to the Office of Radioactive Waste Management, the goal is to submit the license application to the NRC no later than 2008. The NRC then will undertake its own extensive review and testing which is expected to take up to five years. After the license is approved construction on the repository can begin. However, the earliest the repository is now expected to be ready to accept highly radioactive waste is 2017 at the earliest (see Figure 4.2 – Timeline for Opening Yucca Mountain Repository). (OCRWM, 2006: 1) Because of the delay in opening the national repository, inventories of highly radioactive waste continued to mount across the country. In an effort to ensure that Yucca Mountain would be able to handle the increased levels of nuclear waste, the Bush Administration proposed in April 2006 to raise the cap beyond the 70,000 metric ton limit.

As Of July 19, 2006

Design for License Application Complete	**30 November 2007**
Licensing Support Network Certification	**21 December 2007**
Supplemental Environmental Impact Statement (EIS) Issued	**30 May 2008**
Final License Application Verifications Complete	**30 May 2008**
Final Rail Alignment EIS Issued	**30 June 2008**
License Application Submittal	**30 June 2008**
License Application Docketed by NRC	**30 September 2008**

Best-Achievable Repository Construction Schedule

Start Nevada Rail Construction	**5 October 2009**
Construction Authorization from NRC	**30 September 2011**
"Receive and Possess" License Application Submittal to NRC	**29 March 2013**
Rail Access In-Service	**30 June 2014**
Construction Complete for Initial Operations	**30 March 2016**
Start up and Pre-Op Testing Complete	**31 December 2016**
Begin Receipt	**31 March 2017**

Figure 4.2 Timeline for Opening Yucca Mountain Repository

Source: OCRWM, "License Application," Yucca Mountain Project [http://www.ocrwm.doe. gov/ym_repository/about_project/index.shtml]

The Yucca Mountain Project

The highly radioactive waste destined for the Yucca Mountain repository comes from five main sources:

- commercial nuclear power generation (over 50,000 metric tons have been produced)
- as a byproduct of nuclear weapons production (about 2,500 metric tons)
- surplus plutonium from weapons production (about 50 metric tons)
- operational U.S. naval vessels
- research reactors. (OCRWM, 2004b: 6-7)

The planned repository will be capable of storing 70,000 metric tons of spent nuclear fuel and high-level waste.

Yucca Mountain is located approximately 100 miles northwest of Las Vegas adjacent to the Nevada Test Site (see Figure 4.3 – Location of Yucca Mountain). The planned repository comprises about 230 square miles in an arid region near the Armargosa Valley (see Figure 4.4 – Arid Location). The repository will be approximately 1,000 feet below the surface in solid rock (volcanic tuff) and 600-1,600 feet above the water table (see Figure 4.5 – Repository Concept).

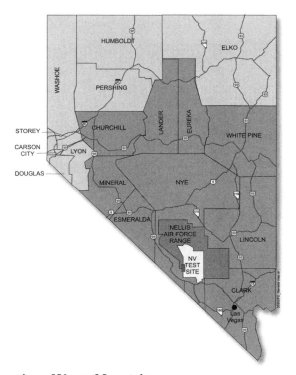

Figure 4.3 Location of Yucca Mountain
Source: OCRWM. Yucca Mountain Project [http://www.ocrwm.doe.gov/ymp/about/quickfacts.shtml]

Figure 4.4 Arid Location
Source: OCRWM. Yucca Mountain Project [http://www.ocrwm.doe.gov/ymp/about/remote.shtml]

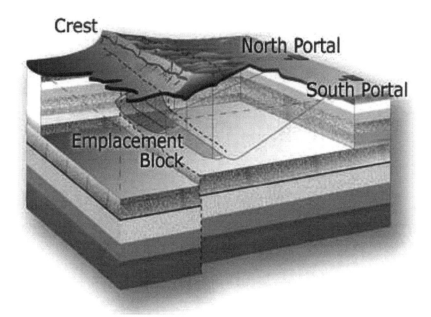

Figure 4.5 Repository Concept
Source: OCRWM. Yucca Mountain Project [http://www.ocrwm.doe.gov/ymp/about/concept.shtml]

Federal regulations mandate that a national repository for highly radioactive waste must have at least one natural and one engineered barrier to protect against water seepage that could reach the waste container, degrade it, and act as transport mechanism for any radioactive material that my leak from the disposal container. (OCRWM, 2004a: 24a) To achieve this objective, the design concept for the Yucca Mountain repository is based on five basic principles (see Figure 4.6 – Proposed Repository Design):

- allow only minimal water seepage into the repository
- limit the amount of radioactive material released from the repository by use of engineered barriers
- use the site's natural barriers to dilute any radioactive material that ultimately is released from the repository
- ensure the waste canisters contain the radioactive material for an extended period
- allow only a small amount of leakage of radioactive material due to corrosion. (NWTRB, 1996: 5)

Figure 4.6 Proposed Repository Design
Source: OCRWM, Program Briefing, "The Proposed Design for a Repository at Yucca Mountain" [http://www.ocrwm.doe.gov/ymp/about/concept.shtml]

First, because of the arid climate, DOE expects a minimal seepage of water into the underground repository. On average, only 7.5 inches of precipitation occurs annually at the site. Because of the dry conditions, the vast majority of the rainfall is expected either to evaporate at the surface or run off. Thus, only a small amount of water is expected to seep into the repository as long as the region remains arid.

Second, the waste canisters are designed to provide containment of the radioactive material for an extended period. The canisters consist of two thick metal containers: an inner stainless steel cylinder and an outer corrosion resistant nickel alloy cylinder. The double cylinder system is expected to contain the radioactive waste for a long period of time.

Third, engineering features are designed to limit corrosion which should reduce the amount of radioactive material that is expected to leak (see Figure 4.7 – Engineered Features). For example, a metal corrosion resistant drip shield will cover the waste canisters to protect against falling debris and water.

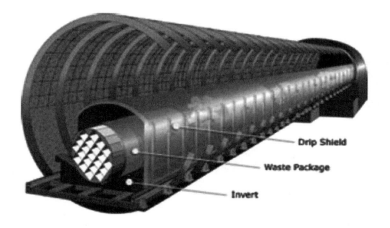

Figure 4.7 Engineered Features
Source: OCRWM, Yucca Mountain Project [http://www.ocrwm.doe.gov/ymp/about/efeatures.shtml]

Fourth, since the waste canisters are expected to be subjected to some corrosion, engineering barriers are designed to limit any release of radioactive material that may breach the waste container. For example, the waste canisters will sit on an "invert" (steel frame filled with crushed volcanic rock) which is designed to slow the movement of any radioactive particles that may leak out of the waste containers onto the rock.

Finally, the site's natural barriers and remote location are expected to provide a substantial dilution of any radioactive material that may leak from the repository. The solid rock is expected to slow the transport of any radioactive material that may leak from the waste canisters and the remote location of the repository will minimize the impact on human populations of any radioactive material that is released from the repository. (OCRWM, 2005: 1)

Thus, according to DOE, the repository design and engineered barriers are expected to geologically isolate the highly radioactive waste for thousands of years. Nevertheless, the intense heat and radioactivity associated with the highly radioactive waste are expected eventually to breach the waste containers. Thus, in spite of the geological and engineered barriers, there is considerable controversy over the efficacy of the repository design to be able to fully isolate the highly radioactive waste for thousands of years.

Advantages and Disadvantages of Permanent Disposal

There are a number of advantages and disadvantages associated with building a national repository for highly radioactive waste. For example, there are technical, societal, security and political benefits and costs associated with finding a permanent solution to the problem of what to do the highly radioactive waste that continues to accumulate at sites across the country (see Table 4.3 – Advantages and Disadvantages of Permanent Disposal).

Table 4.3 Advantages and Disadvantages of Permanent Disposal

Advantages

- improve homeland security by reducing the most vulnerable waste
- improved energy security by avoiding the premature shutdown of commercial nuclear reactors
- nuclear-generated power does not contribute to global warming
- economic benefit since on-site storage is expensive
- societal benefits such as not deferring problem to future generations (e.g., safety, health, and economic costs)

Disadvantages

- geologic and environmental uncertainties about the Yucca site
- societal costs of perceived fairness and equity for Nevada
- politicized and contentious

Advantages

There are a number of potential benefits associated with establishing a national repository for disposing of spent nuclear fuel from commercial nuclear power facilities and high-level waste from defense activities. The most important benefit concerns homeland security. As former Energy Secretary Spencer Abraham pointed out, there are a number of homeland security advantages associated with a long-term solution of what to do with highly radioactive waste. (Abraham, 2002: 2-3) First, a permanent repository is needed to provide a "pathway" for radioactive waste from the Navy's nuclear powered fleet maintaining the viability of the fleet. Second, a repository would help promote non-proliferation objectives by disposing of weapons grade plutonium from decommissioned weapons. Third, the continued accumulation of radioactive waste across the country near millions of Americans presents a concern for homeland security with respect to future terrorism.

The second advantage concerns energy security. Establishing a permanent repository would secure a long-term pathway for spent nuclear fuel from commercial nuclear power facilities. This, in turn, would prevent the premature shutdown of reactors due to the lack of sufficient on-site storage space. Since nuclear power accounts for about one-fifth of the total U.S. electricity generation, any reduction in nuclear generated power would make the U.S. more dependent on foreign sources of energy. Moreover, as the current facilities near the end of their life expectancy, there will be increased pressure for the construction of additional nuclear generated electric power sources to off-set the expected loss in nuclear generated electricity. If additional nuclear power facilities are not constructed, then either fossil fuel plants or alternative energy sources (e.g., geothermal, solar, wind, tidal) will need to be brought on-line to prevent any significant reduction in electricity generation nationally.

A third advantage involves environmental concerns such as global warming. Since fossil fuel fired power generation has substantial environmental costs (e.g., air pollution, acid deposition, and carbon dioxide contribution to global warming) there likely will be opposition to building additional coal and oil burning plants. While alternative energy sources (e.g., solar, geothermal) may be more environmentally sound, there still are unresolved technical and cost-effectiveness problems. Thus, policy makers will find that difficult choices will be required to ensure sufficient electricity for the future. Conversely, nuclear power does not generate either pollution or carbon dioxide; it is (with the exception of highly radioactive waste) more environmentally sound than other more traditional energy sources. Generating more electricity from nuclear power, rather then coal or oil fired power plants would help the U.S. to lower its pollution and carbon dioxide emissions. Under these circumstances, the nuclear power option may appear more attractive which will add pressure to resolve the nuclear waste storage problem.

Fourth, there are also a number of economic incentives to establish a national disposal facility for highly radioactive waste in Nevada. First, under the current storage regime, the nuclear power industry is forced to spend increased resources to store spent nuclear fuel on-site. As suitable storage space runs out, more expensive storage methods (e.g., dry-cask storage) have been needed to cope with the problem. The federal government could be liable for substantial damages for its failure to adhere to the 1998 deadline to accept the highly radioactive waste. Second, the Nuclear Waste Fund (established by the 1982 NWPA–Title III, Sec. 302) was created to pay for costs associated with the permanent storage of highly radioactive waste and is funded by assessing utilities one-mil (i.e., one-tenth of one cent) per kilowatt hour of nuclear generated electricity. This charge ultimately increases the cost of electricity for consumers. The Fund already has accumulated several billion dollars and is projected to disperse over $50 billion during its total system life-cycle costs. The utilities are continuing to bear the brunt of current on-site storage costs and point out that although they have been charged with funding the operational costs of disposing of the nuclear waste, they are still saddled with the on-site storage costs. As the Fund continues to accumulate assets, there will be increased pressure to open a national repository to spend the money for its intended purpose. Third, nuclear utilities have sued DOE for not taking title to the spent nuclear fuel in 1998 as mandated by the 1982 NWPA. Thus, the longer the period before a permanent repository is ready to accept the nuclear waste, the larger the monetary judgment against the federal government likely will be.

Finally, addressing the issue of what to do with the spent nuclear fuel generated by commercial nuclear power utilities and the high-level defense wastes now ensures that the problem is not deferred to future generations. This will prevent future generations from being saddled with the safety, health and economic costs of coping with the highly radioactive waste generated in earlier decades.

Disadvantages

Concerns over the fairness of the site selection process, as well as the geologic and environmental suitability of the site, have intensified conflict. Thus, societal (e.g., perceived fairness and equity) and technical (e.g., geologic and environmental problems) concerns have led to increased politicization of the issue.

Many observers have criticized the site selection process. Some members of Congress have openly admitted that the 1987 legislation was driven by political, not technical or environmental considerations. Representative Al Swift (D-WA) reportedly stated that the process of selecting Yucca Mountain as the national repository for highly radioactive waste was purely political, and summed up the process by saying, "I am participating in a nonscientific process – sticking it to Nevada." (Erickson, 1994: 36) And even Senator J. Bennett Johnston (D-LA), sponsor of the 1987 legislation to single-out Yucca Mountain, conceded that Nevada was victimized by "a cruel trick on Christmas Eve: the waste and no goodies." (Shapiro 1988, 264) In addition to Congress, the Director of DOE's Civilian Radioactive Waste Management Program also commented on the political nature of the process and stated that the selection of Yucca Mountain "slam-dunked Nevada." (Suplee, 1995: A18) It is clear that the 1987 Act violated provisions of the original legislation which mandated that several locations would be studied in order to find the most suitable site. Congress chose Nevada, over that state's objections, undoubtedly because Nevada had a small population and limited political influence. Clearly, Nevada had less political power than the other two finalists (Texas or Washington) which had larger and more influential congressional delegations.

In addition to concerns over the site selection process, there are a number of geologic problems associated with the region which could make it potentially unsuitable for the permanent storage of highly radioactive waste. For example, there are concerns about volcanic disturbances, potential earthquakes, rapid groundwater movement and hydrothermal activity at the site. The mountain was formed millions of years ago by a series of explosive volcanic eruptions that deposited ash and material that compressed together to create layers of rock. While the explosive type of volcano is extinct in the Yucca Mountain area, scientists are studying seven small, dormant volcanoes in the area. Two of these volcanoes may have been active within the last 10,000 years, a relatively short time in geologic terms. (DOE, 1992: 9) The area also contains over 30 known earthquake fault lines. In 1992, an earthquake measuring 5.6 on the Richter Scale caused considerable damage to the Yucca Mountain project field operations center. There also are concerns about the movement of groundwater which could transport any radioactive material that may leak from the disposal facility. Some tests indicate a possible flow rate in excess of what is considered desirable, prompting further testing. (DOE, 1997: ES 8-11) Moreover, a number of scientists (a former DOE scientist, as well as scientists working for the State of Nevada) believe that there is renewed evidence of hydrothermal activity at the site. Percolated water (geothermal water that has risen to the surface and drained back down) was found in a borehole 1600 feet deep as late as March 1995. Similar pockets of percolated water were found in 1993 and 1994 at similar depths; as well as in the 1980s, before the present studies of Yucca Mountain began. (DOE, 1995: 1-2) It is important to note that one of the former DOE hydrology scientists resigned from DOE because he viewed the Yucca Mountain project as environmentally unsafe. (Lowry, 1993: 32 and Rosenbaum, 1992: 10) The issue has become even more contentious in 2005 since it was revealed that USGS personnel may have submitted fraudulent water flow data.

In contrast, DOE scientists currently studying Yucca Mountain believe that, at this point, none of the previously stated problems give cause for declaring the site unsuitable for the permanent disposal of highly radioactive waste. First, they assert that the probability of a volcano erupting in the region over the next 10,000 years is very remote – calculated to be about 1 in 70 million per year (DOE, 1996, 1) Second, they believe that the risk of damage to underground facilities from faults in the region is small because the amount of movement on local faults has been small, with possibly many thousands of years between movements. Moreover, experience with earthquakes throughout the world has shown that underground structures can withstand the ground motion generated by earthquakes better than aboveground structures. Additionally, in actual nuclear weapons tests at the nearby Nevada Test Site, mine tunnels have withstood ground motion from underground nuclear explosions that are greater than any ground motion anticipated at the Yucca Mountain site. (DOE, 1996: 8) Finally, while there is some concern about the rate of groundwater movement, testing is being conducted to measure the scope of the problem. According to DOE, "The water may flow down through a fault or crack, stopping before it reaches the repository horizon..." (DOE, 1996: 134) Moreover, engineering barriers such as drip shields have been added in order to provide some protection from water movement.

The problem of scientific uncertainty over the geologic suitability of the area for the disposal of highly radioactive waste has intensified the political conflict. Ever since Yucca Mountain was first designated as a possible repository, Nevada has been highly critical of the site selection process. While enhanced citizen awareness over the perceived costs associated with radioactive waste disposal (potential economic, environmental, health and safety problems) has spawned an active and vibrant "not in my backyard" (NIMBY) movement in the states designated as a potential nuclear waste repository, opposition has been especially intense in Nevada. Many Nevadans believe that the political and economic incentives influencing the federal government and other proponents of the Yucca Mountain project are too great to allow for an objective evaluation. Nevada is concerned that the time and energy spent on the site study, as well as the investment of billions dollars, has created momentum and expectations to proceed with the project in spite of its flaws. In addition, according to Nevada, the credibility of some DOE scientists is questionable since some work for employee profit-sharing companies that are working on the project. (State of Nevada, 2004a: 1) The longer these companies work on the highly radioactive waste storage project, the more these scientists profit financially. In addition to potential safety problems, Nevada is concerned that a radioactive waste dump could stigmatize Nevada as a "nuclear waste state" which could adversely impact tourism – the main source of revenue for the state. Public opinion polls commissioned by the state have shown that this concern may be justified.[10] Compounding Nevada's concern about being singled out to site the only permanent repository for highly radioactive waste is DOE's proposal to turn part of the Nevada Test Site into "the largest low-level nuclear waste disposal site in the nation." (State of Nevada, 2004c: 2)

Summary

In the final analysis, the politics involved in establishing a national repository for the disposal of highly radioactive waste likely will intensify in the future. In spite of political and geologic concerns, the nuclear power industry and DOE undoubtedly will continue to pressure the federal government to locate the permanent disposal facility at Yucca Mountain since any effort to remove the site from consideration will only serve to delay finding a solution to the problem of disposing of spent nuclear fuel and high-level waste.

The road to establishing a national repository for spent nuclear fuel and high-level waste has been long and treacherous. As previously stated, there are a number of compelling reasons for establishing a national repository for spent nuclear fuel and high-level waste. Clearly, homeland security and energy security concerns are real. Moreover, there are economic incentives and political pressures to build the repository as soon as possible.

On the other hand, there are equity concerns that must be addressed in order to minimize political conflict. As pointed out by government officials:

In situations in which distributional equity is impossible to obtain, the fairness of decision procedures becomes of paramount concern to members of the public. (NAS, 2001: 79)

Clearly, this was not the case with the Yucca Mountain decision. The procedures establishing a permanent disposal facility outlined in the 1982 NWPA were violated by the subsequent 1987 NWPA Amendments Act (PL 100-203). While the 1982 Act mandated that multiple sites would be considered and sites would be constructed in both the East and West, the 1987 Act singled out Yucca Mountain as the only site to

---

10 See State of Nevada, "State of Nevada Socioeconomic Studies: Biannual Report, 1993-95," State of Nevada Nuclear Waste Project Office, p. 9. Also see series of studies released by Nevada's Nuclear Waste Project Office (available on-line at their website) concerning the adverse socioeconomic impact of a nuclear waste disposal facility.

be studied. Thus, it is not surprising that Nevada has refused to accept the fairness of the site selection process.

The State of Nevada, environmental groups and the grassroots NIMBY movement will continue to actively oppose the present siting and approval process. At this time, for a number of reasons, it would be difficult for Nevada's politicians to support the Yucca Mountain project. First, due to the intense opposition within the state, it would be political suicide to support establishing a nuclear waste disposal facility in Nevada. Second, concern over the potential negative impact on tourism provides a strong incentive for continued opposition to establishing any nuclear waste repository in the state. Moreover, environmental groups will likely continue to oppose the current siting process due to concerns over potential geologic instability (volcanic activity, earthquakes, ground water movement and hydrothermal activity). Finally, due to the political, economic, and environmental concerns outlined above, the grassroots NIMBY movement likely will remain active and vibrant. This will ensure continued pressure on Nevada's lawmakers to oppose any nuclear waste disposal facility.

The inability to reach a solution to the problem of what to do with the spent nuclear fuel from commercial nuclear power plants and high-level waste from defense facilities has intensified conflict and has heavily politicized the issue. As a result, the Yucca Mountain nuclear waste repository will not be ready before 2017 at the earliest – more likely not until 2020. In spite of legislative attempts to resolve the impasse, and all the litigation designed to force DOE to take responsibility for the nuclear waste, the nuclear utilities are still responsible for storing their spent nuclear fuel on-site (either in cooling pools or in above-ground casks).

The political reality is that a permanent geological repository for highly radioactive waste is quite contentious both at the state and national levels. The political conflict has at least delayed the project, and possibly could ultimately derail the proposed repository. However, a permanent repository would provide very good homeland security since the nuclear waste would be stored below ground in a highly secure facility. Thus, there would be a significantly reduced risk of a terrorist attack or sabotage. Moreover, energy security would be excellent since a long-term pathway for spent nuclear fuel would be available, eliminating the need to prematurely shut down nuclear power reactors due to a lack of sufficient on-site storage space (see Figure 4.8 – Permanent Radioactive Waste Disposal: Homeland Security and Energy Security Risk v. Political Risk) for a summary of the overall risk analysis associated with the permanent disposal of highly radioactive waste.

When a permanent repository is constructed, large quantities of highly radioactive waste will have to be transported across the country. Chapter 5 discusses the transportation of radioactive waste by providing an overview of transportation issues, examines safety concerns such as the potential for sabotage or terrorist attack, and the potential for an accident. Transportation issues such as the integrity of the transportation casks, the transportation routes, and emergency response in the event of a release of radiological material also are discussed.

	Poor	Fair	Good	Excellent
Homeland Security Risk			X*	
Energy Security Risk				X
	Low	Moderate	High	
Political Risk			X	

Figure 4.8 Permanent Radioactive Waste Disposal: *Homeland Security and Energy Security Risk v Political Risk*
* Denotes "very good" homeland security

Key Terms

ocean/sub seabed disposal
1972 London Convention on the Prevention of Marine Pollution by Dumping of Wastes and Other Matter
extraterrestrial disposal
remote uninhabited island disposal
ice sheet disposal
deep bore hole disposal
reprocessing (or partitioning)
proliferation
transmutation
deep borehole disposal
geologic disposal
NIMBY

Useful Web Sources

Las Vegas Review Journal [http://www.reviewjournal.com]
State of Nevada, Nuclear Waste Project Office [http://www.state.nv.us/nucwaste]
U.S. Congressional Legislation [http://thomas.loc.gov]
U.S. Department of Energy [http://www.doe.gov]
U.S. Department of Energy, Yucca Mountain Repository [http://www.ocrwm.doe.gov/ym_repository/index.shtml]
U.S. Department of Energy, Office of Civilian Radioactive Waste Management [http://www.ocrwm.doe.gov]
U.S. Nuclear Waste Technical Review Board [http://www.nwtrb.gov]

References

(Abraham, 2002) Abraham, Spencer. Letter to President Bush recommending approval of Yucca Mountain for the development of a nuclear waste repository, Secretary of Energy, U.S. Department of Energy, 14 February 2002.

(Abrahms, 2005) Abrahms, Doug. "As Yucca project stalls, Utah nuke waste dump hits fast track", *Reno Gazette Journal*, 4 April 2005 [http://www.rgj.com/news/stories.html/2005/04/02/96157.php]

(BMU, 2005) *Environmental Policy: Joint Convention on the Safety of Spent Fuel Management and on the Safety of Radioactive Waste Management*, Report Under the Joint Convention by the Government of the Federal Republic of Germany for the Second Review Meeting in May 2006, Federal Ministry for the Environment, Nature Conservation and Nuclear Safety, September 2005 [http://www.bmu.de/files/english/nuclear_safety/application/pdf/2nationaler_bericht_atomenergie_en. pdf]

(Bodman, 2005) Bodman, Samuel. "Statement From Secretary of Energy, Samuel Bodman," >*energy.gov*, 16 March 2005 [http://www.energy.gov/engine/content.do?PUBLIC_ID=17629&BT_CODE=PR_PRESSRELEASES&TT_CODE=PRESSRELEASE]

(Bodman, 2007) Bodman, Samuel W. Letter to the Honorable Richard B. Cheney supporting the "Nuclear Fuel Management and Disposal Act." March 6, 2007 (with enclosures) [http://www.ocrwm.doe.gov/info_library/newsroom/documents/2007bill.pdf]

(Bush, 2002) Bush, George W. "Presidential Letter to Congress," 15 February 2002 [http://www.whitehouse.gov/news/releases/2002/02/20020215-10.html]

(Carter and Pigford, 2004) Carter, Louis J. and Thomas H. Pigford "A Better Way at Yucca Mountain," Preliminary Report of Work in Progress, Meeting of the National Research Council's Board on Radioactive Waste Management, 20 September 2004.

(CBS News, 2005) "EPA Revises Nuclear Dump Standards," 9 August 2005 [http://www.cbsnews.com/stories/2005/08/09/tech/main768203.shtml?CMP=ILC-SearchStories]

(CNR, 2005) *Canadian National Report for the Joint Convention on the Safety of Spent Fuel Management and on the Safety of Radioactive Waste Management, Second Report*, October 2005 [http://www.nuclearsafety.gc.ca/pubs_catalogue/uploads/2005_joint_convention_report_English.pdf]

(Courier, 1996) "Court: U.S. must take nuclear waste by 1998," *The Courier*, 24 July 1996: A1.

(Defra, 2006) *The United Kingdom's Second National Report on Compliance with the Obligations of the Joint Convention on the Safety of Spent Fuel Management and on the Safety of Radioactive Waste Management*, Department for Environment, Food and Rural Affairs, February 2006 [http://www.defra.gov.uk/environment/radioactivity/government/international/pdf/jointconreport06.pdf]

(DOE, 1992) *DOE's Yucca Mountain Studies*, Yucca Mountain Site Characterization Project, Office of Civilian Radioactive Waste Management, DOE/RW-0345P, December 1992.

(DOE, 1995) Yucca Mountain Project Studies, "Drillers find more perched water in Yucca Mountain's unsaturated zone," *Of Mountains & Science*, Summer 1995, [http://www.ymp.gov/ref_shlf/ofms/default.html].

(DOE, 1996a) Yucca Mountain Project Studies, "Damage to FOC from '92 quake less than first cited," *Of Mountains & Science*, Winter 1996: 129-40.

(DOE, 1996b) Yucca Mountain Project Studies, "Project scientists assess tritium traces for evidence of fast routes to repository," *Of Mountains & Science*, Winter 1996: 134, 139.

(DOE, 1996c) Yucca Mountain Project Studies, "Yucca Mountain Volcanic Hazard Analysis Completed," 24 September 1996 [http://www.ymp.gov/wha_news/ymppr/pvha.htm].

(DOE, 1997) "Site Characterization Progress Report: Yucca Mountain, Nevada," U.S. Department of Energy, DOE/RW-0496, No. 15, April 1997.

(DOE, 2006) "Technical Report Confirms Reliability of Yucca Mountain Technical Work," 17 February 2006 [http://www.energy.gov/news/3220.htm]

(DOE, 2007) "DOE to Send Proposed Yucca Mountain Legislation to Congress," Office of Public Affairs, U.S. Department of Energy, March 6, 2007 [http://www.ocrwm.doe.gov/info_library/newsroom/documents/Yucca_leg_03_06_07_press_release.pdf]

(EIA, 2001) "World Cumulative Spent Fuel Projections by Region and Country," Energy Information Agency, U.S. Department of Energy, 1 May 2001 [http://www.eia.doe.gov/cneaf/nuclear/page/forecast/cumfuel.html]

(ENR, 1998) "High Court Leaves DOE on the Hook," *Engineering News Record*, 30 November/7 December 1998.

(EOP, 1997) "Statement of Administration Policy: S. 104 – Nuclear Waste Policy Act of 1997," Executive Office of the President, 7 April 1997 [http://clinton3.nara.gov/OMB/legislative/sap/105-1/S104-s.html]

(Erickson, 1994) Erickson, Kai. "Out of Sight, Out of Our Minds," *The New York Times Magazine*, 6 March 1994: 36-41, 50, 63.

(FAEA, 2006) *The National Report of the Russian Federation on Compliance with the Obligations of the Joint Convention on the Safety of Spent Fuel Management and the Safety of Radioactive Waste Management*, Prepared for the Second Review Meeting in Frames of the Joint Convention on the Safety of Spent Fuel Management and the Safety of Radioactive Waste Management, Federal Atomic Energy Agency, Moscow 2006 [http://www-ns.iaea.org/downloads/rw/conventions/russian-federation-national-report.pdf]

(FANC, 2006) *Second Meeting of the Contracting Parties to the Joint Convention on the Safety of Spent Fuel Management and on the Safety of Radioactive Waste Management*, Federal Agency for Nuclear Control, Kingdom of Belgium, National Report, May 2006 [http://www.avnuclear.be/avn/JointConv2006.pdf]

(GAO, 2001) "Nuclear Waste: Technical, Schedule, and Cost Uncertainties of the Yucca Mountain Repository Project," Government Accounting Office, GAO-02-191, December 2001 [http://www.gao.gov/new.items/d02191.pdf]

(Guinn, 2002) Guinn, Kenny C. "Official Notice of Disapproval of the Yucca Mountain Site," State of Nevada, Office of the Governor, 8 April 2002.

(Highfield, 2005) Highfield, Roger. "Seabed scars show power of ocean quake," *The UK Telegraph On-Line*, 10 February 2005 [http://telegraph.co.uk/news/main. jhtml;sessionid=FZFDIIAMDUEKDQFIQMFCM5OAVCBQYJVC?xml=/ news/2005/02/10/wsea10.xml&secureRefresh=true&_requestid=59553]

(Holt, 1996) Holt, Mark. "Civilian Nuclear Waste Disposal," *CRS Issue Brief*, Congressional Research Service, 21 November 1996 [http://www.cnie.org/nle/ waste-2.html]

(Holt, 1998) Holt, Mark. "Civilian Nuclear Spent Fuel Temporary Storage Options," Congressional Research Service, 96-212 ENR (Updated 27 March 1998) [http:// cnie.org/NLE/CRSreports/Waste/waste-20.cfm]

(Holt, 2006) Holt, Mark. "Civilian Nuclear Waste Disposal," Congressional Research Service, RL33461 (Updated 19 September 2006) [http:ncseonline.org/ NLE/CRSreports/06Sep/RL33461.pdf]

(Holt and Davis, 1996) Holt, Mark and Zachary Davis, "Nuclear Energy Policy," *CRS Issue Brief*, Congressional Research Service (Updated 5 December 1996) [http://www.cnie.org/nle/eng-5.html]

(HSK, 2005) *Implementation of the Obligations of the Joint Convention on the Safety of Spent Fuel Management and on the Safety of Radioactive Waste Management*, Second National Report of Switzerland in Accordance with Article 32 of the Convention, Department of Environment, Transport, Energy and Communications, September 2005 [http://www.hsk.ch/english/files/pdf/joint-convention_CH_sept-05.pdf]

(IAEA, 2006a) *Nuclear Power and Sustainable Development*, International Atomic Energy Agency, April 2006 [http://www.iaea.org/Publications/Booklets/ Development/npsd0506.pdf]

(IAEA, 2006b) *Nuclear Technology Review 2006*, International Atomic Energy Agency, IAEA/NTR/2006, August 2006 [http://www.iaea.org/OurWork/ST/NE/ Pess/assets/ntr2006.pdf]

(Kraft, 1996) Kraft, Michael E. "Democratic Dialogue and Acceptable Risks: The Politics of High-Level Nuclear Waste Disposal in the United States," in *Siting by Choice: Waste Facilities, NIMBY, and Volunteer Communities*, Don Munton, ed., (Georgetown University Press: Washington DC, 1996), pp. 108-41.

(London, 1993) London Convention on the Prevention of Marine Pollution by Dumping of Wastes and Other Matter, 1972, Annexes I and II, amended in 1993 [http://www.londonconvention.org]

(Loux, 1995) Loux, Robert R. "Nuclear Waste Policy at the Crossroads," [http:// www.astro.com/yucca].

(Lowry, 1993) Lowry, David. "2010: America's Nuclear Waste Odyssey," *New Scientist*, 6 March 1993: 30-3.

(LVRJ, 2002) "Yucca Mountain: Guinn vetoes Bush," *Las Vegas Review-Journal*, 9 April 2002 [http:www.lvrj.com/cgi-bin/printable.cgi?/lvrj_home/2002/Apr-09-Tue-2002/news/18476447]

(LVRJ, 2005a) "E-mails say scientists fabricated quality assurance on Yucca Mountain," *Las Vegas Review Journal*, 2 April 2005 [http:www.reviewjournal. com/lvrj_home/2005/Apr-02-Sat-2005/news/26204008.html]

(LVRJ, 2005b) "Energy officials turn shy in talk about Yucca Schedule," *Las Vegas Review Journal*, 11 March 2005 [http:www.reviewjournal.com/lvrj_home/2005/Mar-11-Fri-2005/news/26048040.html]

(Macfarlane, 2001) Macfarlane, Alison. "Interim Storage of Spent Fuel in the United States," *Annual Review of Energy and the Environment*, vol. 26, 2001: 201-35.

(METI, 2005) *Joint Convention on the Safety of Spent Fuel Management and on the Safety of Radioactive Waste Management*, National Report of Japan for the Second Review Meeting, October 2005 [http://www.meti.go.jp/english/report/index.html]

(MITYC, 2005) *Joint Convention on the Safety of Spent Fuel Management and on the Safety of Radioactive Waste Management*, Second Spanish National Report, October 2005 [http://www.mityc.es/NR/rdonlyres/BF3E47F5-7861-4A28-8FB3-8F2DBAB183A3/0/ConvencionInforme2_ing.pdf]

(NAS, 1957) "The Disposal of Radioactive Waste On Land," Report of the Committee on Waste Disposal of the Division of Earth Sciences, National Academy of Sciences, National Research Council, Publication 519, September 1957 [http://www.nap.edu/books/NI000379/html]

(NAS, 2001) "Disposition of High-Level Waste and Spent Nuclear Fuel: The Continuing Societal and Technical Challenges," Committee on Disposition of High-Level Radioactive Waste Through Geologic Isolation, Board on Radioactive Waste Management, Division on Earth and Life Sciences, National Academy of Sciences, National Research Council, National Academy Press, Washington D.C., 2001 [http://www.nap.edu/openbook/0309073170/html/R1.html]

(NEA, 2006) *Safety of Geological Disposal of High-level and Long-lived Radioactive Waste in France*, An International Peer Review of the "Dossier 2005 Argile" Concerning Disposal in the Callovo-Oxfordian Formation, Nuclear Energy Agency, Organisation for Economic Co-operation and Development, NEA No. 6178, OECD 2006 [http://www.nea.fr/html/rwm/reports/2006/nea6178-argile.pdf]

(Nixon, 1995) Nixon, Will. "High Energy," *E/The Environmental Magazine*, May/June 1995 [http://www.adams.ind.net/Environment.html]

(Nuke-Energy, 2002) "OCRWM Chief Says Something: 2010 Milestone is 'Tight'," Nuke-Energy, July 2002 [http://www.nuke-energy.com/data/stories/July02/Chu.htm]

(NWTRB, 1996) "Nuclear Waste Management in the United States," Nuclear Waste Technical Review Board, Topseal Conference, June 1996 [http://www.nwtrb.gov/reports/wastemgt.pdf]

(OCRWM, 2000) "Civilian Radioactive Waste Management Program Plan (Revision 3)," Office of Civilian Radioactive Waste Management, U.S. Department of Energy, February 2000 [http://www.ocrwm.doe.gov/pm/pdf/pprev3.pdf]

(OCRWM, 2001a) "Belgium's Radioactive Waste Management Program," Office of Civilian Radioactive Waste Management, U.S. Department of Energy, DOE/YMP-0407, June 2001 [http://www.ocrwm.doe.gov/factsheets/doeymp0407.shtml]

(OCRWM, 2001b) "Canada's Radioactive Waste Management Program," Office of Civilian Radioactive Waste Management, U.S. Department of Energy, DOE/YMP-0408, June 2001 [http://www.ocrwm.doe.gov/factsheets/doeymp0408.shtml]

(OCRWM, 2001c) "China's Radioactive Waste Management Program," Office of Civilian Radioactive Waste Management, U.S. Department of Energy, DOE/YMP-0409, June 2001 [http://www.ocrwm.doe.gov/factsheets/doeymp0409.shtml]

(OCRWM, 2001d) "Finland's Radioactive Waste Management Program," Office of Civilian Radioactive Waste Management, U.S. Department of Energy, DOE/YMP-0410, June 2001, available at: [http://www.ocrwm.doe.gov/factsheets/doeymp0410.shtml]

(OCRWM, 2001e) "France's Radioactive Waste Management Program," Office of Civilian Radioactive Waste Management, U.S. Department of Energy, DOE/YMP-0411, June 2001 [http://www.ocrwm.doe.gov/factsheets/doeymp0411.shtml]

(OCRWM, 2001f) "Japan's Radioactive Waste Management Program," Office of Civilian Radioactive Waste Management, U.S. Department of Energy, DOE/YMP-0413, June 2001 [http://www.ocrwm.doe.gov/factsheets/doeymp0413.shtml]

(OCRWM, 2001g) "Russia's Radioactive Waste Management Program," Office of Civilian Radioactive Waste Management, U.S. Department of Energy, DOE/YMP-0413, June 2001 [http://www.ocrwm.doe.gov/factsheets/doeymp0413.shtml]

(OCRWM, 2001h) "Spain's Radioactive Waste Management Program," Office of Civilian Radioactive Waste Management, U.S. Department of Energy, DOE/YMP-0415, June 2001 [http://www.ocrwm.doe.gov/factsheets/doeymp0415.shtml]

(OCRWM, 2001i) "Switzerland's Radioactive Waste Management Program," Office of Civilian Radioactive Waste Management, U.S. Department of Energy, DOE/YMP-0417, June 2001 [http://www.ocrwm.doe.gov/factsheets/doeymp0417.shtml]

(OCRWM, 2001j) "The United Kingdom's Radioactive Waste Management Program," Office of Civilian Radioactive Waste Management, U.S. Department of Energy, DOE/YMP-0418, June 2001 [http://www.ocrwm.doe.gov/factsheets/doeymp0418.shtml]

(OCRWM, 2001k) "Radioactive waste: an international concern," Office of Civilian Radioactive Waste Management, U.S. Department of Energy, DOE/YMP-0405, June 2001 [http://www.ocrwm.doe.gov/factsheets/doeymp0405.shtml]

(OCRWM, 2004a) "Assessing the Future Safety of a Repository at Yucca Mountain," Office of Civilian Radioactive Waste Management, U.S. Department of Energy, CD-ROM, 1 March 2004.

(OCRWM, 2004b) "Annual Report to Congress – December 2004,"Office of Civilian Radioactive Waste Management, U.S. Department of Energy, DOE/RW-0569 [http://www.ocrwm.doe.gov/pm/program_docs/annualreports/04ar/fy_2004.pdf]

(OCRWM, 2004c) "Oklo: Natural Nuclear Reactors," Office of Civilian Radioactive Waste Management, U.S. Department of Energy, DOE/YMP-0010, November 2004 [http://www.ocrwm.doe.gov/factsheets/doeymp0010.shtml]

(OCRWM, 2005) "Disposal Options: Reprocessing and Transmutation," Office of Civilian Radioactive Waste Management, U.S. Department of Energy [http://ocrwm.doe.gov/ymp/about/reprocess.shtml]

(OCRWM, 2007) "Repository Sites Considered in the United States," Yucca Mountain Project, Office of Civilian Radioactive Waste Management, U.S. Department of Energy [http://www.ocrwm.doe.gov/ym_repository/about_project/sitesconsidered.shtml]

(Ouchida, 1996) Ouchida, Kurt. "The Yucca Mountain Saga: A Project in Decay," [http:www.astro.com/yucca/html].

(Pulsiper, 1993) Pulsiper, Allan G. "The Risk of Interim Storage of High Level Nuclear Waste," *Energy Policy*, July 1993: 798-812.

(Rissmiller, 1996) Rissmiller, Kent. "Equality of Status, Inequality of Result: State Power and High-Level Radioactive Waste," *Publius: The Journal of Federalism*, Winter 1993: 103-18.

(Rosenbaum, 1992) Rosenbaum, David B. "Yucca Mountain Draws Fire Even From Some within DOE," *Engineering News Report*, 27 April 1992: 10.

(Shapiro, 1989) Shapiro, Fred C. "Yucca Mountain," *The New Yorker*, May 22, 1988, in *Taking Sides: Clashing Views on Controversial Issues*, 3rd. ed., Theodore D. Goldfarb, ed. (Dushkin Publishing Group, Inc.: Guilford, CT, 1989), pp. 260-69.

(State of Nevada, 2004a), "Concerns About Yucca Mountain," State of Nevada Nuclear Waste Project Office [http://www.well.com/user/rscime/yucca/concerns.html].

(State of Nevada, 2004b), "State of Nevada Socioeconomic Studies: Biannual Report, 1993-1995," State of Nevada Nuclear Waste Project Office [http://www.state.nv.us/nucwaste/yucca/sebian95.html].

(State of Nevada, 2004c), "Why Nevada is Opposed to Yucca Mountain," State of Nevada Nuclear Waste Project Office [http://www.well.com/user/rscime/yucca/wyop1.html]

(STUK, 2005) *Joint Convention on the Safety of Spent Fuel Management and on the Safety of Radioactive Waste Management*, 2nd Finnish National Report as referred to in Article 32 of the Convention, STUK-B-YTO 243, October 2005 [http://www.stuk.fi/julkaisut/stuk-b/stuk-b-yto243.pdf]

(Suplee, 1995) Suplee, Curt. "A Nuclear Problem Keeps Growing," *The Washington Post*, 31 December 1995: A1, A18, A19.

(URL, 2006) "Underground Research Laboratory," Applied Seismology Laboratory, Liverpool University Department of Earth Sciences [http://www.liv.ac.uk/seismic/research/url/url.html]

(USCM, 2002) "Resolution: Transportation of High-Level Nuclear Waste," U.S. Conference of Mayors, Adopted at the 70th Annual Conference of Mayors, Madison, WI, 14-18 June 2002.

(U.S. Congress, S. 1271), "Nuclear Waste Policy Act of 1996," *Congressional Record*, 104th Congress, 2d Session.

(U.S. Congress, S. 1936), "Nuclear Waste Policy Act of 1996," *Congressional Record*, 104th Congress, 2d Session.

(U.S. Congress, S. 104), "Nuclear Waste Policy Act of 1997," *Congressional Record*, 105th Congress, 1st Session.

(U.S. Congress, 2002a) "Final Vote Results for Roll Call 133," H.J. 87 [http://clerkweb.house.gov/cgi-bin/vote.exe?year=2002&rollnumber=133]

(U.S. Congress, 2002b) "U.S. Senate Roll Call Votes," S.J. 34, 107th Congress, 2nd Session [http://www.senate.gov/legislative/vote1072/vote_00167.html]

Chapter 5

Transportation

To date, the radioactive waste debate generally has revolved around either the construction of a centralized permanent national repository at Yucca Mountain, Nevada; or, to a lesser extent, the proposed construction of an interim storage site (e.g., Skull Valley, Utah, or Owl Creek, Wyoming). However, more recently the focus of opposition has shifted to the transport of the highly radioactive waste to any potential centralized facility. Concerns over the safety and security of transporting large amounts of highly radioactive waste over long distances, sometimes through heavily populated areas, have sparked significant opposition to any potential transportation plan. In fact, opponents of any disposal or storage option now view the transportation issue as the next line of defense against building and licensing any proposed centralized facility.

This chapter discusses the transportation of highly radioactive waste. First, an overview of radioactive waste transportation is given. Second, transportation safety concerns are outlined. Third, cask integrity, transport routes, and the efficacy of emergency response in the event of an incident with a release of radioactive material are discussed.

The Transportation Debate

National security and energy security realities will force policy makers to move more assertively in finding a solution – most likely a centralized repository or an interim storage facility – to the nuclear waste storage and disposal issue. However, any centralized storage or disposal option will necessitate moving large quantities of highly radioactive waste across long distances from existing nuclear facilities to a more secure facility. This will, in turn, generate substantial opposition to any proposed transportation initiative since many of the same concerns voiced over the vulnerability of the radioactive waste on-site can be made for the radioactive waste in transit.

What Radioactive Materials are Transported?

A variety of radioactive materials are transported on a daily basis across the country (e.g., uranium ore for processing, nuclear fuel rods for commercial power plants, radioisotopes for medical purposes and research, nuclear defense materials and radioactive waste).[1] According to DOE, "about 3 million packages of radioactive materials are shipped each year in the United States." (DOE, 1999a: 1) Radioactive materials are transported by ground (rail and highway), water (ship and barge) and air. The type and use of the radioactive material generally determines the mode of transport. For example, almost two-thirds of the radioactive materials shipped each year are radioisotopes for medical purposes. (DOE, 1999a: 5) Since this material is short-lived and must be delivered to hospitals and medical facilities quickly, it generally is shipped by air. Usually, this material has a low level of radioactivity and does not prompt much concern. Spent nuclear fuel, on the other hand, is more problematical due to its high-level of radioactivity and generally is shipped by either rail or highway transport. While radioactive waste can be shipped by barge, less than 1 percent of radioactive waste has been shipped by water since it is a slow method of transport and is limited geographically to those reactors that are on a body of water. No commercial spent nuclear fuel has been shipped by barge in the United States. (DOE, 1999a: 10)

Radioactive materials are shipped in specially packaged government approved containers for transport.[2] Design specifications, which become more rigorous as the radiation levels of the content increases, require the packaging to withstand an accident without the release of the radioactive contents. Four general types of packaging are used: *excepted packaging* is used to transport materials with extremely low levels of radioactivity; *industrial packaging* (IP-1, IP-2, IP-3) is used to transport materials which present a limited hazard; *Type A packaging* is used to transport radioactive materials with higher concentrations of radioactivity; *Type B packaging* is used to transport materials with the highest radioactive levels. (DOE, 1999a: 14-15) Due to the limited hazard associated with radioactive material shipped in excepted packaging, only general design requirements must be met. Industrial packaging will vary according to the type of radioactive material being shipped. For example, IP-1 packaging must meet the same general design requirements for excepted packaging, IP-2 packaging must pass free-drop and stacking tests, and IP-3 packaging must also pass water spray and penetration tests. Type A packaging must be able to withstand a series of tests (penetration, vibration, compression, free drop, and water spray) designed to determine the ability of the container to survive an accident without the release of radioactive material. Type B packaging must pass even a more rigorous sequence of tests due to the increased radioactivity of its cargo (free drop – from 30 feet, puncture – 40 inch drop onto a steel rod, heat – exposure to 1,475 degrees for 30 minutes, immersion – immersed under 3 feet of water, and crush-dropping an 1,100 pound mass from 30 feet onto the container). (DOE, 1999a: 16-18) (See Figure

---

1   The Transportation Resource Exchange Center (T-REX) is an excellent source of information for the transport of radioactive materials. See (T-REX, 2003a) [http://www.trex-center.org].

2   See (T-REX, 2003c) "Spent Nuclear Fuel Transportation Regulations," Transportation Resource Exchange Center, ATR Institute, University of New Mexico, Updated 13 March 2003 [http://www.trex-center.org/caskregs.asp] for a good resource on radioactive materials shipping regulations.

5.1 – For Certification, Casks Must Pass Following Tests). Additional crash tests designed to replicate highway and rail accidents also have been conducted.

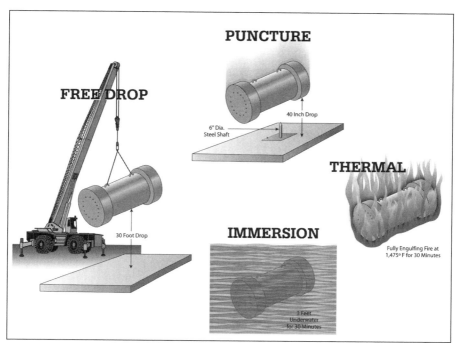

Figure 5.1 For Certification, Casks Must Pass Following Tests
Source: "Spent Nuclear Fuel Transportation," OCRWM [http://www.ocrwm.doe.gov/wat/pdf/snf_trans.pdf]

For highly radioactive spent fuel, specially designed Type-B *rail and highway transportation casks* that have been certified by the NRC are used (see Figure 5.2 – Rail Transportation Cask and Figure 5.3 – Highway Transportation Cask). Shipment by rail is the preferred alternative since there is a smaller risk of accident and more material can be shipped at one time. For highway transport, highly radioactive spent fuel must be identified as a *Highway Route Controlled Quantity (HRCQ) shipment*, which means that certain provisions must be met. First, all shipments must be clearly marked with the appropriate radioactive material warning sign (see Figure 5.4–Radioactive Waste Transportation Symbol). Second, "preferred routing" must be used which takes into account factors such as the potential accident rate along the transit route, the transit time, population density affected by the route, and time of day and day of the week. (DOE, 1999a: 20)

Figure 5.2 Rail Transportation Cask
Source: OCRWM Program Briefing, "Transportation," OCRWM [http://www.ocrwm.doe.
gov/pm/programbrief/briefing.htm]

Figure 5.3 Highway Transportation Cask
Source: OCRWM Program Briefing, "Transportation," OCRWM [http://www.ocrwm.doe.
gov/pm/programbrief/briefing.htm]

Figure 5.4 Radioactive Waste Transportation Symbol
Source: "Transporting Spent Nuclear Fuel and High-level Radioactive Waste to a National Repository: Answers to Frequently Asked Questions," OCRWM [http://www.ocrwm.doe.gov/wat/pdf/snf_transfaqs.pdf]

There are a variety of federal and state actors involved in regulating the transport of highly radioactive waste. The myriad of regulatory actors overlap responsibilities in monitoring the safety of the nuclear material during transport.

The Federal Bureaucracy   The Department of Energy is an important entity that regulates the transport of highly radioactive waste since it is tasked to bring together a wide variety of organizations that have an interest in the transportation of radioactive waste materials through its National Transportation Program (NTP). The NTP mission is to provide policy guidance, technical and management support, and operational services to assure the availability of safe and secure transportation of non-classified DOE nuclear materials. (NTP, 2005: 1)

The Nuclear Regulatory Commission is responsible for regulating the performance of packaging and transport operations of shippers of spent nuclear fuel and high-level radioactive waste. (DOE, 2002a: 4)   The Office of Nuclear Security and Incident Response is responsible for developing emergency preparedness programs to respond to a variety of nuclear emergencies.

The Environmental Protection Agency also provides guidance and training to other federal and state agencies in preparing for emergencies at U.S. nuclear plants, and transportation accidents involving shipments of radioactive materials.   The Federal Radiological Emergency Response Plan (FRERP) designates the EPA as the lead federal agency for emergencies involving radiological material not licensed or owned by a federal agency or an agreement state. (FRERP, 1996: II-2)

The Department of Transportation (DOT) has the primary responsibility for establishing and enforcing standards for the shipment of highly radioactive waste. The DOT's Federal Motor Carrier Safety Administration (FMCSA) enforces radiological materials regulations such as the classification of the materials and proper packaging to ensure the safe and secure transportation by highway. (FMCSA, 2005) DOT's Pipeline and Hazardous Materials Safety Administration (PHMSA) oversees the safety of highway hazardous materials shipments including radiological waste. (PHMSA, 2004b) The PHMSA's Office of Hazardous Materials Safety oversees the Hazardous Materials Transportation Program, which is tasked to identify and manage risks presented by the transportation of hazardous material such as radioactive waste. (PHMSA, 2005b) DOT's Federal Railroad Administration (FRA) is the government agency responsible for monitoring highly radioactive waste shipments by rail. (DOE, 2002a: 4) The FRA has developed a safety compliance plan to promote the safety of highly radioactive waste by rail. (FRA, 1998: iv)

The Federal Emergency Management Agency (FEMA) also plays a role in the transportation of nuclear waste. FEMA would play a crucial role in the response and recovery phase of any transportation accident involving nuclear waste that would result in the release of radioactivity.

The Courts   The conflict caused by the politicization of the highly radioactive waste issue has increased litigation. As a result, the federal courts have played an increasingly active role in more recent years and have ruled in a variety of cases involving radioactive waste. More recently, litigation has focused on the safety of the transportation of radioactive waste and opponents of building a centralized storage/ disposal facility now view the transportation issue as a primary means to block the construction of any centralized facility.

State/Local Governments   The states and local governments clearly can play an important role in the radioactive waste transportation debate since they are involved with the transportation of highly radioactive waste through their jurisdictions. Moreover, the federal government permits state and local governments to pass legislation that specify requirements for transportation of radioactive materials within their jurisdictions as long as these laws are consistent with federal law, and do not impede interstate commerce. (DOE, 1999a: 23) Almost 500 laws and regulations concerning radioactive waste have been passed by the states. Most of these laws principally have affected radioactive waste highway transport such as restrictions on routes, registration requirements, vehicle restrictions, and permits that affect the transportation of spent nuclear fuel. (Reed, 2000: 9)

While some states such as Nevada, Utah and South Carolina have attempted to block radioactive waste shipments, others such as Illinois and New Mexico have been proactive in regulating the transport of nuclear waste. Illinois has developed a model transportation program since it was required to accept spent nuclear fuel from out of state reactor operators that had concluded agreements with its reprocessing facility at Morris. New Mexico also has adopted a proactive approach, due to the construction of the Waste Isolation Pilot Plant (WIPP) near Carlsbad, which accepts transuranic waste from out-of-state defense industry facilities. Officials from both states have opted to pass regulations that would ensure a more active role in monitoring and regulating the storage and transport of the radioactive waste within state boundaries.

The primary concern for local governments has been the emergency response capability to any potential accident involving the transport of radioactive materials. Local governments have attempted to influence policy by passing resolutions calling for the federal government to undertake certain actions. In 2002, the U.S.

Conference of Mayors passed a resolution opposing the transportation of highly radioactive waste across state borders until the federal government can provide "adequate funding, training and equipment to protect public health and safety" for all cities along the proposed transportation route. (USCM, 2002: 1) In 2003, the National Association of Counties adopted a policy resolution that urged DOE to develop a transportation plan and accompanying environmental impact statement for highly radioactive waste shipments. (NACO, 2002: 25-26)

Interest Groups   There are a number of state and local government affiliated organizations that have lobbied intensively on radioactive waste transportation issues. For example, as discussed previously, the Western Governors' Association has issued several resolutions regarding spent fuel transportation and the National Association of Counties and the U.S. Mayors Conference have adopted policy statements calling for more training and funding for local emergency response along the proposed transportation routes for the spent nuclear fuel. In addition, the Western Interstate Energy Board (an organization of western states and three Canadian provinces) has been critical of DOE and has reiterated the western governors' positions on the issues. (WIEB, 1996: 2; WIEB 1998: 1; and WIEB 2002:1) The National Association of Regulatory Utility Commissioners (NARUC), a non-profit organization whose members include government agencies engaged in regulating utilities and carriers in the fifty states, District of Columbia, Puerto Rico and the Virgin Islands also has issued a number of resolutions regarding highly radioactive waste.

There also have been a number of citizen and industry interest groups active in the nuclear waste transport debate. The Transportation Safety Coalition has been especially active in attempting to block the Yucca Mountain project. Some of the environmental and citizen activist opposition to either the construction of a centralized storage or disposal facility, or the transport of highly radioactive waste, is genuine due to safety and health concerns. Other opposition groups, however, hope to eventually shut down the commercial nuclear power industry by preventing the transport of the spent nuclear fuel which would force the shutdown of reactors due to the limited availability of on-site storage space. In contrast, the commercial nuclear power industry and supporters understand that there will be an insufficient amount of on-site spent fuel storage capability in the near future and that a number of reactors will be forced to shut down if the on-site waste is not transported to either a centralized interim storage facility or a permanent disposal facility. The Nuclear Energy Institute (NEI), which is the nuclear power industry's main lobbying organization, has been especially vocal and active in support of building a centralized storage or permanent disposal facility and has stressed the positive safety record of transporting radioactive materials over the years. Storage and transportation cask vendors also have been active in the radioactive waste transportation debate, usually focusing on the sturdiness of the casks.

The Media   While the media has not yet focused much attention on the highly radioactive waste transportation question, they could play an important future role in the debate. For example, unfair critical reporting by the media could serve to undermine public confidence in the safety of transporting the radioactive waste to a storage or disposal facility. This, in turn, would likely increase citizen opposition to transportation plan. On the other hand, if the media focuses on the positive safety record and the low likelihood of an accident that could result in a release of radiological material, then citizen opposition will likely be minimized. Clearly, any major incident or accident involving the transport of highly radioactive waste will receive extensive media coverage at the local, state and national levels. And, if a significant amount of radioactivity were released, then the "Three-Mile Island/ Chernobyl effect" would likely dominate the news and prompt a flurry of negative reporting on the safety of transporting highly radioactive waste across the United States.

The Concern over Safety

There has been much discussion and controversy over the safety of transporting large amounts of radioactive waste over long distances sometimes through heavily populated areas. Concern primarily focuses on four areas: the potential for sabotage or a terrorist attack, the potential for an accident resulting in the release of radioactive material, the ability of the transportation casks to maintain their integrity in the event of an incident, and the distance and location of the rail and highway transportation routes.[3]

The Potential for Sabotage or Terrorist Attack

Opponents of transporting radioactive waste cite the potential for sabotage or a terrorist attack on the shipment as another reason why the waste should be left on-site. For example, the Western Governors Association (WGA) passed a resolution in 1998 calling on the NRC to take steps to protect the shipments against a terrorist attack. (NCSL, 1998: 7) The WGA issued another policy resolution ("Assessing the Risks of Terrorism and Sabotage Against High-Level Nuclear Waste Shipments to a Geologic Repository or Interim Storage Facility,") in 2001, prior to the terrorist attack on 9/11, which questioned the vulnerability of spent fuel shipments to sabotage and terrorism. The resolution requests the NRC to undertake a comprehensive reassessment of terrorism and sabotage risks for radioactive waste shipments and incorporate risk management and countermeasures "in all DOE transportation plans relating to operation of a repository, interim storage facility, and/or intermodal transfer facility." (WGA, 2001: 2-3) Moreover, a study commissioned by the State of Nevada criticized the federal government's preparation and readiness for a possible terrorist or sabotage incident and argued that a high-energy explosive could breach a cask dispersing radiological material. (Halstead, no date: 1)

---

3   See (T-REX, 2000) "A Consideration of Routes and Modes in the Transportation of Radioactive Materials and Waste: An Annotated Bibliography," Transportation Resource Exchange Center, ATR Institute, University of New Mexico, 31 December 2000 [http:// www.trex-center.org/dox/routesbib.pdf] for an excellent resource of articles available on the transportation of radioactive waste question.

In spite of these concerns, however, a number of government-sponsored studies have concluded that the potential for a significant release of radiation as a result of a sabotage or terrorist attack on spent fuel transportation containers is relatively low. For example, a 1977 Sandia study concluded that a successful sabotage attack on a spent fuel container would not result in any immediate deaths and fewer than 100 latent cancer deaths from the released radioactive material. A 1983 Sandia report found that an armor penetrating missile attack on a mock spent fuel assembly would only release a miniscule number of particles that could be inhaled. Finally, a 1999 Sandia report also concluded that an armor-piercing missile attack on a shipping container would not likely result in a major release of radioactive material. (GAO, 2003: 37-39)

Clearly, the events of 9/11 have further underscored the need for adequate security for highly radioactive waste shipments. Government officials respond that all necessary provisions for protecting the shipments have been taken: escorts for all spent fuel shipments, 24-hour monitoring of the shipment, safeguarding schedules, and coordinating with local law enforcement agencies. (DOE, 2002a: 5) The commercial nuclear industry asserts that spent fuel is not an attractive target since the transportation casks are so heavy (75-100 tons), the spent fuel cannot be used for a nuclear device without reprocessing, any attempt to penetrate the cask would result in an unacceptable level of exposure to radiation, the transportation casks are proven to be sturdy, and any potential damage from a breach in a cask would be minor compared to many other targets. (PFS, 2003: 5) According to the U.S. Department of Energy, "Should an act of sabotage be attempted, the consequences are expected to be minor due to the protective nature of the cask." (DOE, 1999a: 27) Nevertheless, there has been concern exhibited over the security and safety of highly radioactive waste shipments. The release of radiological material from a sabotage or terrorist attack involving a spent nuclear fuel shipment will be the issue that opponents use as the main point of their opposition to transporting highly radioactive waste across the country.

The Potential for an Accident

The potential for an accident and subsequent release of radioactive materials has generated considerable opposition to transporting large amounts of highly radioactive waste. The federal government touts its safety record for past DOE radioactive materials shipments and argues that the risk for a serious accident that would result in a significant release of radioactive material is low. Critics, on the other hand, argue that there have been a significantly higher number of "incidents" that have resulted in a release of radioactive material. Moreover, transportation opponents point out that any centralized storage or disposal facility would entail shipping vast amounts of radioactive waste across great distances, which would significantly increase the likelihood of a serious accident.

According to DOE, the "nation's nuclear transportation record is impressive." (DOE, 1999a: 2) The federal government boasts that its safety record is unparalleled:

Over the last *30 years*, there have been over *2,700 shipments* of spent nuclear fuel traveling over *1.6 million miles* in the United States. There has never been a release of radioactive materials harmful to the public or environment – not one. (OCRWM, 2002c: 2, emphasis original)

This statement does not imply, however, that there have been no accidents resulting in the release of radioactive material, just that no accident resulted in a release that the government regarded as "harmful to the environment." In fact, DOE has acknowledged that a "small number of shipments have been involved in traffic accidents." (DOE, 2002a: 4) From January 1971 to December 1998 there were 8 documented accidents involving shipments of spent nuclear fuel. (DOE, 1999a: 28)

It is difficult to project the exact number of spent nuclear fuel shipments to the Yucca Mountain facility, and thus the risk to the public. DOE has published different statistics on the number of anticipated radioactive waste shipments in different publications. For example, in some publications, DOE estimates that approximately 175 shipments will take place annually if much of the waste is transported on dedicated trains loaded with several spent fuel transportation casks.[4] Thus, over a 24-year period there would be approximately 3,200 shipments by rail and 1,100 shipments by highway to the Yucca mountain repository. (DOE, 2002a: 3; and DOE, 2002b: 9) However, in the final environmental impact statement for Yucca Mountain, DOE states that a "mostly legal-weight truck scenario would involve about 53,000 shipments (2,200 annually), and the mostly rail scenario would involve approximately 10,700 shipments (450 annually)." (DOE, 2002b: S-68, 69) Clearly, some of the difference in the anticipated number of shipments is a result of whether a mostly rail or mostly truck scenario is used (DOE has not fully determined yet whether the spent fuel will be transported primarily by rail or highway). However, the discrepancy in the number of anticipated rail shipments (i.e., 3,200 v. 10,700) also is due to whether "casks" or "shipments" are counted (e.g., multiple transportation casks could be loaded on one rail shipment). It is interesting to note that the lower transportation figures are included in the publications that are intended for public consumption (i.e., widely available on the web), whereas the higher figures are used in the environmental impact statement which is a long and tedious technical document to read; and thus, will be read by far fewer people.

Using the wide disparity in these figures to calculate the risk of an accident will provide fodder for both proponents and opponents of the Yucca Mountain project. Proponents (e.g., the federal government and the commercial nuclear industry) will emphasize the small number of expected shipments (i.e., 4,300), while the opponents

---

4   DOE announced in April 2004 that rail would be the "preferred mode" of transport. Subsequently, DOE also announced that it would use only "dedicated" trains to ship the nuclear waste. (OCRWM, 2005: 1) Using the rail only scenario would reduce the required number of shipments. Using dedicated trains would increase safety since no other potentially flammable or harmful materials would be included with the shipment. However, in spite of the earlier announcement that rail would be the preferred mode of transport, in 2006, DOE stated that both rail and truck (annually about 130 rail and 45 truck) shipments will be used to transport nuclear waste to Yucca Mountain. (OCRWM, 2006: 9)

(selected state/local governmental organizations, and environmental/citizen activist groups) will cite the higher number of shipments (i.e., 55,000) since a larger number of shipments will equate to a greater chance for an accident. For example, if the approximately 2,700 documented spent nuclear fuel shipments that took place over the past 30-years experienced 8 accidents, then approximately .3 percent of the spent fuel shipments were involved in accidents. If this accident rate (.3 percent) is applied to the DOE forecast of approximately 4,300 rail and highway shipments of radioactive waste to Yucca Mountain, then as many as 12.9 accidents involving spent nuclear fuel can be expected. However, if the 10,700-shipment scenario is used, and the .3 percent accident rate is applied, then 32.1 accidents could be expected. On the other hand, if the 55,000- shipment scenario is used, applying the .3 percent accident rate would yield an expected 165 accidents.

There have been a number of studies conducted over the years on the potential accident rate of transporting radioactive waste. A 1987 Lawrence Livermore National Laboratory federally funded study on the potential for a nuclear waste transportation accident (generally referred to as the Modal Study) predicted a .2 percent risk for a possible accident resulting in a release of radioactive material with "severe consequences." (Holt, 1998: 6) Applying the .2 percent anticipated "severe consequence" accident rate to the possible 4,300 shipments of spent nuclear fuel to the Yucca Mountain repository would result in 8.6 accidents with the potential for a significant release of radiological material. If the .2 percent risk is applied to the 10,700 mostly rail shipment scenario and the 55,000 mostly truck scenario, then there could be 21.4 and 110 accidents respectively resulting in a "severe consequences" release of radioactive material. More recently, a 2000 study by Sandia National Laboratory predicted a .007 percent truck and .004 percent train accident risk that might cause the release of radioactive material. (GAO, 2003: 11) Applying these expected accident rates to the anticipated number of shipments would yield .3 truck and 3.9 rail accidents that could result in the release of radioactive material.

Opponents of transporting highly radioactive waste are more critical of the government's safety record and use different criteria for determining the spent fuel safety transportation record. While DOE states there were 8 spent fuel "accidents" from 1971-1998, opponents cite (using a 1996 DOE report as the source) 72 "incidents" from 1949-1996.[5] (Nevada, 1996: 1-4) An *incident* would be any reported problem that occurred during any point of the shipment phase, not just a rail/vehicle *accident* that occurred during transit. According to Public Citizen, a watchdog public interest group, "incidents" can be significant. To support their claim, they have cited incidents (not accidents) involving transportation casks that resulted in a significant release of radioactive material. According to Public Citizen, in 1980 and 1981, two separate "incidents" involved transportation casks that became so severely contaminated that they emitted radiation levels far beyond allowed maximum limits. (Public Citizen, 2002: 1-2)

---

5   See (Public Citizen, 2002) "A Brief History of Irradiated Nuclear Fuel Shipments: Atomic Waste Transport 'Incidents' and Accidents the Nuclear Power Industry Doesn't Want You to Know About," Public Citizen [http://www.citizen.org/print_article.cfm?ID=5873] for a summary of the 1996 DOE report.

The accident/incident terminology difference impacts on public trust and confidence in the context of evaluating the accident risk for spent fuel shipments. First, DOE admits the "carriers (not law enforcement officials) are required to report hazardous materials incidents to the U.S. Department of Transportation (DOT)." (DOE, 1999b: J-68) Second, the NRC requires the compilation of reports involving loss, exposure to, or release of radioactive materials. Data maintained by Sandia National Laboratory in the *Radioactive Materials Incident Report (RMIR) database*, is categorized and defined differently and is "not consistent with the definitions used in other (DOT) databases." (DOE, 1999b: J-69) DOE recognizes radioactive materials as a subset of the general term hazardous materials. Accordingly, transportation regulations contain no distinction between an "accident" and an "incident," and "incident" is the term used to describe a situation that must be reported. (DOE, 1999b: J-68 and J 69) Subsequently, due to the variation in database languages DOE has chosen "not to use RMIR to estimate transportation accident rates." (DOE, 1999b: J-69)

Critics also assert that over the past 35 years there have been only a relatively small number of spent nuclear fuel shipments each year, and those shipments traveled a relatively short distance usually crossing only one to three states. (Public Citizen, 2002: 2) They argue that if a centralized storage/disposal facility is built, there will be a significant increase in the number of highly radioactive materials shipments across many states. They assert that this will, in turn, significantly raise the possibility of an accident with a serious release of radiological material.

Transportation Cask Integrity

There has been considerable focus on how well a transportation cask could survive an incident without releasing radioactive material.[6] In fact, according to the commercial nuclear power industry:

> The safety of SNF [spent nuclear fuel] shipments with respect to radiological impacts, especially in the event of a transportation accident, *is ensured, in large measure by the casks that contain the SNF.* (NRC, 2001: 5-1, emphasis added)

While the federal government maintains that no "harmful" release of radiological materials has occurred during the shipment of spent nuclear fuel, there has been considerable concern exhibited over the integrity of the transportation casks (see Figure 5.5 – Generic Rail Cask for Spent Fuel and Figure 5.6 – Generic Highway Cask for Spent Fuel). In response, there have been attempts to discern how well a transportation cask could withstand a serious accident. For example, the Livermore Modal Study focused on transportation risks of the radioactive material by calculating the *strain* (i.e., stress or damage) on a cask under different accident scenarios. There are three levels of "strain." *Level 1* is where the strain on the cask's inner wall does

---

6   See (NRC, 2003a) "Safety of Spent Fuel Transportation," U.S. Nuclear Regulatory Commission, NUREG/BR-0292, March 2003 [http://www.nrc.gov/reading-rm/doc-collections/nuregs/brochures/br0292] for a good overview on transportation casks. Also see (T-REX, 2003b) "Spent Nuclear Fuel Transportation Cask Tests" [http://www.trex-center.org/casktest.asp] for a good resource on transportation cask testing.

not exceed .2 percent (the limit that stainless steel can withstand without violating its integrity). A *Level 2* strain is .2 percent-2 percent (stainless steel does not regain its shape and lead shielding may separate which may allow radiation to leak). *Level 3* strain is between 2 percent-30 percent (can cause large distortions of the metal shape and the cracking of welded areas which may result in a serious release of radioactive material). (Holt, 1998: 5-6)

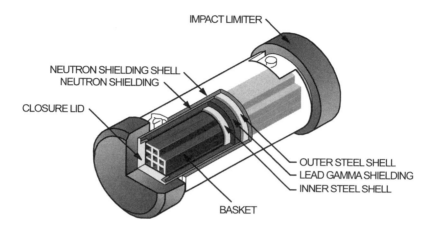

Figure 5.5 Generic Rail Cask for Spent Fuel
Source: NRC (2005) "Transporting Spent Nuclear Fuel," Diagram and Description of Typical Spent Fuel Transportation Cask, U.S. Nuclear Regulatory Commission [http://www.nrc.gov/waste/spent-fuel-storage/diagram-typical-trans-cask-system.doc]

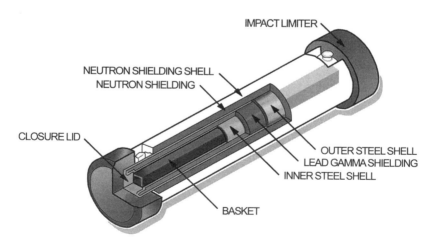

Figure 5.6 Generic Highway Cask for Spent Fuel
Source: NRC (2005) "Transporting Spent Nuclear Fuel," Diagram and Description of Typical Spent Fuel Transportation Cask, U.S. Nuclear Regulatory Commission [http://www.nrc.gov/waste/spent-fuel-storage/diagram-typical-trans-cask-system.doc]

The Modal Study also looked at several serious transportation accidents not involving radiological materials and attempted to determine if the transportation casks would have been able to withstand the accident scenarios without a breach of the container. In most cases, it was determined that the cask would have been able to withstand the accident with essentially no damage. However, transportation opponents assert that a train derailment or serious highway accident that involved a serious prolonged fire could result in a significant release of radiological material. (Holt, 1998: 6) For example, the *2001 Baltimore Tunnel accident* (which resulted in a major fire that lasted for four days) is cited by transportation opponents as a classic example of an accident that could result in a significant release of radiological material. They assert that an accident of this magnitude could result in a serious release of radioactive material since the fire lasted far beyond the test limits.

Most evidence, however, would suggest that the transportation casks would be able to withstand most accident scenarios. For example, a 2005 NRC study indicated that a fire similar to the Baltimore Tunnel incident would not result in a release of radioactive material. (Struglinski, 2005: 1) Additional tests conducted by Sandia National Laboratories in 2001 came to a similar conclusion. Sandia conducted a series of crash tests involving transportation casks, none of which would have resulted in a release of radioactive material. Four tests were conducted: a truck carrying a 22 ton cask was crashed into a wall at 60 mph, and then again at 84 mph; crashing a locomotive traveling at 81 mph into a 25 ton cask on a truck trailer; and crashing a rail car with a 74 ton spent fuel cask into a wall and then engulfing cask and rail car in a jet-fuel fire for 90 minutes. (Holt, 1998: 7) The report issued on the Sandia tests focused on five topic areas: (1) package performance during a collision, (2) package performance during a fire, (3) spent nuclear fuel behavior during an accident, (4) highway and railway accident conditions and probabilities, and (5) other transportation safety issues. (Sprung, 2001: ix)

Nevertheless, previous tests have been criticized by the State of Nevada as being too optimistic since they assert that some damage to the casks could have resulted in the possible release of radiological material. (Holt, 1998: 7) Furthermore, opponents argue that previous transportation cask crash tests are not necessarily representative of the most severe potential accident scenario and assert that casks "are not designed to withstand all credible highway and rail accidents." (Lamb, 2001: i) They further argue that even a small radiological release could adversely affect a large number of individuals if the accident occurred in a populated area.

The Transportation Routes

Any centralized disposal or storage plan will necessitate transporting large amounts of highly radioactive waste across the country. And, as the maps of the rail and highway transportation routes for spent nuclear fuel reveal, substantial amounts of highly radioactive waste clearly will pass through many communities, some with large populations (see Figure 5.7 – Rail Transportation Routes and 5.8 – Highway Transportation Routes). As would be expected, the federal government (NRC and DOT) has the ultimate authority to monitor and control radioactive waste shipments. The federal government is required to notify the states, in advance, of an impending spent nuclear fuel shipment through their jurisdiction.[7] In addition, according to DOT regulations, states have the authority to designate alternate routes in order to minimize the risk to the public. However, the states do not have the authority to refuse shipment of the material as evidenced by the standoff between the federal government and South Carolina. In 2002, South Carolina attempted to block shipments of highly radioactive waste into the state. The governor vowed to prevent the radioactive waste from entering the state even going as far as using the state police for roadblocks. (Jordan, 2002: 1) Ultimately, the state lost the dispute as the principle of national supremacy prevailed.

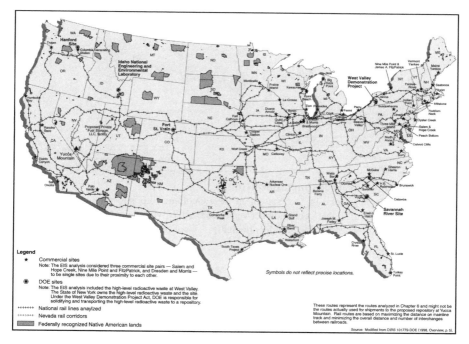

Figure 5.7 Rail Transportation Routes

Source: Final Environmental Impact Statement for a Geologic Repository for the Disposal of Spent Nuclear Fuel and High-Level Radioactive Waste at Yucca Mountain, Nye County, Nevada, OCRWM [http://www.ocrwm.doe.gov/documents/feis_2/vol_1/ch_06/index1_6.htm]

---

7   While DOE notifies states in advance of spent fuel shipments, there is no notification for classified shipments or routine low-level hazardous materials shipments. Moreover, local governments are not formally notified by the federal government; local notices are at the discretion of the states. (NTP, 2000: 3)

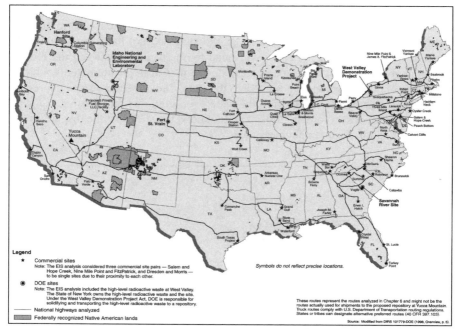

Figure 5.8 Highway Transportation Routes
Source: Final Environmental Impact Statement for a Geologic Repository for the Disposal of
Spent Nuclear Fuel and High-Level Radioactive Waste at Yucca Mountain, Nye County, Nevada,
OCRWM [http://www.ocrwm.doe.gov/documents/feis_2/vol_1/ch_06/index1_6.htm]

In the Event of a Release of Radiological Material

Although most transportation studies indicate that the risk for a serious release of
radiological material is low, there is no guarantee that radioactive material will not
be released. In fact, DOT has been criticized by the General Accounting Office
(GAO) for its lack of preparation for the impending radioactive waste shipments.
After a thorough review of DOT radioactive waste transportation plans, a GAO
inspector reportedly stated, "At this time DOT is not fully prepared for the forecasted
increase in shipments." (Tetreault, 2002: 1) The findings of the GAO report are
not comforting, since in the final analysis, it is the ability to respond quickly and
efficiently to any radiological materials accident that is crucial to mitigating the
potential effects on the public.

    In the event of a radiological accident, local emergency response personnel will
be the first on-scene responders. DOE boasts that over 1200 emergency responders
have been trained and about $20 million in emergency assistance funding has been
provided to 23 states for training for a radiological accident. (OCRWM, 2002c:
7) However, while some assistance has been provided by DOE to support their
shipments (i.e., shipments to or from DOE facilities), provisions outlined in NWPA
specifically state that DOE will provide both "technical assistance and funds" (out

of the Nuclear Waste Fund) to states prior to beginning spent nuclear fuel shipping to a repository.[8] (NWPA, 1982: Section 180c) This NWPA (Section 180c) issue has become a dominant theme among groups representing both local and state governments anticipating the effects of the proposed actions to transport spent nuclear fuel to Nevada or any other centralized radioactive waste storage facility.

A number of groups such as the Western Governors Association, Western Interstate Energy Board, National Conference of State Legislatures, National Association of Counties, and the Conference of Radiation Control Program Directors have called for the federal government to provide training and funding assistance to train local responders years prior to beginning the transport of spent nuclear fuel to a centralized radioactive waste storage/disposal facility. Congress, under pressure by the state and local governments, has attempted to meet the spirit of NWPA (Section 180c) in past sessions by emphasizing that there would be no shipments of spent nuclear fuel "unless technical assistance and funds to implement procedures for safe transportation and for dealing with emergency situations have been available for a least two years prior to any shipment." (U.S. Congress, 1997: Section 203; U.S. Congress, 2000: Section 201)

State and local governments have the primary responsibility for initially responding to a transportation accident. Thus, the "role of the Federal government in responding to transportation accidents is usually one of supporting the lead role of State, Tribal, and local government." (FEMA, 1988: 11) However, in the event of an accident with a serious release of radioactive material it is unlikely that initial local response personnel will have the capability to contain the situation adequately. The state is tasked with assisting local governments when local emergency response is not adequate for the situation. While the state has better resources to cope with an accident that results in a significant release of radiological material, ultimately it is the federal government that will provide the necessary resources and personnel to manage the situation. The DOE *Radiological Assistance Program (RAP)* was established to assist state and local governments to respond to an accident with a serious release of radioactive materials.[9] A request for assistance would be directed to one of the eight DOE *Regional Coordinating Offices (RCO)*, which could dispatch a *Radiological Assistance Team* to assist in the clean up (see Figure 5.9 – Regional Coordinating Offices). (DOE, 1992: 4) However, the federal government acknowledges that while assistance can be mobilized quickly (as early as four hours), "it may take up to forty-eight hours for [federal assistance] to be fully functional." (TEC/WG, 2000: 10, Federal Register, 1998: 23754) Thus, the state and local government position that there should be no shipment of spent fuel to a centralized facility until technical assistance, funding and training has been provided has considerable merit.

---

8   The 1982 Nuclear Waste Policy Act (as amended) is available on-line at: [http://www.ocrwm.doe.gov/documents/nwpa/css/nwpa.htm]

9   See (DOE, 1992) "Radiological Assistance Program," U.S. Department of Energy, Office of Defense Programs, Order DOE 5530.3, 14 January 1992 [http://www.directives.doe.gov/pdfs/doe/doetext/oldord/5530/o55303c1.pdf] for background on the Radiological Assistance Program.

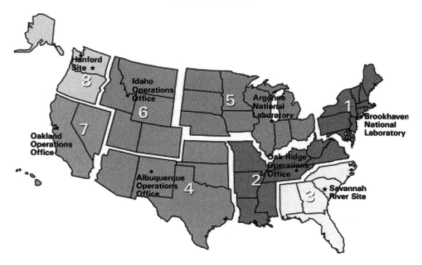

Figure 5.9 Regional Coordinating Offices
Source: "Transporting Spent Nuclear Fuel and High-level Radioactive Waste to a National Repository: Answers to Frequently Asked Questions," OCRWM [http://www.ocrwm.doe. gov/wat/pdf/snf_transfaqs.pdf]

The *Transportation Emergency Preparedness Program (TEPP)*, which is under the auspices of DOE, was created to provide support to federal, state, tribal and local authorities to help them prepare to respond to an incident involving DOE shipments of radioactive material. The Transportation Emergency Preparedness Program integrates transportation and emergency preparedness activities and could serve as a model for the transportation of commercial spent nuclear fuel. The TEPP is implemented on a regional basis using the eight DOE Regional Coordinating Offices that can provide assistance in the event of an accident involving a significant release of radiological material. (TEPP, 2002: 1)

If there is a release of radiation, there will be the issue of compensation to the affected parties. The Price-Anderson Act initially was enacted into law in 1957 and extended to 2025 by the Energy Policy Act of 2005 (P.L. 109-58) to encourage commercial nuclear power production by providing indemnity against losses from nuclear accidents.[10] The Act also provides for the protection against monetary losses due to radiation exposure from an accident during the transport of highly radioactive waste from reactors to a storage or disposal facility.

---

10 See (Callan, 1998) Callan, Joseph L. "NRC's Report to Congress on the Price-Anderson Act," U.S. Nuclear Regulatory Commission, SECY-98-160, 2 July 1998 [http://www.nrc.gov/ reading-rm/doc-collections/commission/secys/1998/secy1998-160/1998-160scy.html] for an excellent discussion of the Price-Anderson Act.

Summary

National security and energy security concerns have prompted the federal government to move more aggressively towards constructing a permanent national repository for highly radioactive waste. Moreover, concerns over the possibility of commercial nuclear reactors having to shut down prematurely due to a lack of adequate on-site storage space for spent nuclear fuel has prompted the nuclear industry to undertake plans for constructing a centralized interim storage facility. Thus, at this time, it appears that it is likely that there will be large amounts of spent nuclear fuel transported across the United States to either the Yucca Mountain repository or to a centralized interim storage facility. In fact, the federal government considers the transport of highly radioactive waste a fait accompli as evidenced by the following quote:

> ...the question is whether, as a national policy, it is best for transportation [of radioactive waste] to be arranged on an ad-hoc basis...or for transportation [of radioactive waste] to be arranged systematically and with years of careful advance planning. (OCRWM, 2002c: 12)

Thus, federal officials believe that the focus should now be on how to transport the radioactive waste safely and efficiently.

In contrast, opponents are not necessarily convinced that the transport of spent nuclear fuel is inevitable. In fact, some view the transportation debate as the last opportunity to block the construction of a centralized facility. Thus, the debate over the safety of transporting large quantities of spent nuclear fuel likely will intensify in the coming years, making the transportation issue the next battleground.

In the final analysis, the locus of decision making on policy, procedures and appropriations in support of any transportation program involving spent nuclear fuel must involve all levels of government (i.e., local, state and federal) and must be collaborative. The political and bureaucratic obstacles to better cooperation and communication between the various levels of government must be overcome in order ensure that the government fulfills its role of protecting public safety. As a former chairman of the National Transportation Safety Board stated:

> History has shown us time and time again that if the essential elements of a safety plan are not put into place before an activity begins, the momentum of the activity overtakes safety considerations. (Hall, 2002: 6)

Thus, it is imperative that a coherent national radioactive waste transportation policy be adopted in the near term in order to provide sufficient time for state and local governments to prepare adequately for any potential future highly radioactive waste shipments.

In the final analysis, security concerns should determine highly radioactive waste policy. However, in reality, political considerations have been the driving force behind radioactive waste policy decisions. Chapter 6 provides a theoretical framework that analyzes the homeland security and energy security risks of on-site storage, centralized interim storage, and permanent disposal, while at the same time factoring in the political ramifications associated with each storage and disposal method.

Key Terms

excepted packaging
industrial packaging
Type A packaging
Type B packaging
rail transportation cask
highway transportation cask
Highway Route Controlled Quantity (HRCQ) shipment
incident
accident
Radioactive Materials Incident Report (RMIR) database
strain
level 1 strain
level 2 strain
level 3 strain
2001 Baltimore Tunnel accident
Radiological Assistance Program (RAP)
Regional Coordinating Offices (RCO)
Radiological Assistance Team
Transportation Emergency Preparedness Program (TEPP)

Useful Web Sources

Nuclear Energy Institute [http://www.nei.org]
Private Fuel Storage, LLC [http://www.privatefuelstorage]
Transportation External Coordination Working Group [http://www.ocrwm.doe.gov/transport/tec.shtml]
Transportation Resource Exchange Center (T-REX) [http://www.trex-center.org]
U.S. Department of Energy [http://www.doe.gov]
U.S. Department of Energy, National Transportation Program [http://www.nnsa.doe.gov/na-26/docs/ntp.pdf]
U.S. Department of Energy, Office of Civilian Radioactive Waste Management [http://www.ocrwm.doe.gov]
U.S. Department of Energy, Transportation External Working Group [http://www.tecworkinggroup.org/]
U.S. Department of Homeland Security, Federal Emergency Management Agency [http://www.fema.gov]
U.S. Department of Transportation, Federal Motor Carrier Safety Administration [http://www.fmcsa.dot.gov]
U.S. Department of Transportation, Federal Railroad Administration [http://www.fra.dot.gov]

U.S. Department of Transportation, Pipeline and Hazardous Materials Safety Administration [http://www.phmsa.dot.gov]

U.S. Environmental Protection Agency [http://www.epa.gov]

U.S. Nuclear Regulatory Commission [http://www.nrc.gov]

U.S. Nuclear Regulatory Commission, Office of Nuclear Material Safety and Safeguards [http://www.nrc.gov/about-nrc/organization/nmssfuncdesc.html]

U.S. Nuclear Waste Technical Review Board [http://www.nwtrb.gov]

Western Governors' Association [http://www.westgov.org]

Western Interstate Energy Board [http://www.westgov.org/wieb]

References

(Abraham, 2002a) Abraham, Spencer. Secretary of Energy Letter to President Bush recommending approval of Yucca Mountain for the development of a nuclear waste repository, U.S. Department of Energy, 14 February 2002.

(Abraham, 2002b) Abraham, Spencer. Secretary of Energy Prepared Witness Testimony, Committee on Energy and Commerce, Subcommittee on Energy and Air Quality, 18 April 2002. [http://energycommerce.house.gov/107/hearings/04182002Hearing505/Abraham859print.htm]

(Callan, 1998) Callan, Joseph L. "NRC's Report to Congress on the Price-Anderson Act," U.S. Nuclear Regulatory Commission, SECY-98-160, 2 July 1998 [http://www.nrc.gov/reading-rm/doc-collections/commission/secys/1998/secy1998-160/1998-160scy.html]

(DOE, 1992) "Radiological Assistance Program," U.S. Department of Energy, Office of Defense Programs, Order DOE 5530.3, 14 January 1992 [http://www.directives.doe.gov/pdfs/doe/doetext/oldord/5530/o55303c1.pdf]

(DOE, 1999a) "Transporting Radioactive Materials: Answers to Your Questions," U.S. Department of Energy, Office of Civilian Radioactive Waste Management, National Transportation Program, June 1999 [http://www.ocrwm.doe.gov/wat/transportation.shtml]

(DOE, 1999b) "Draft Environmental Impact Statement for a Geologic Repository for the Disposal of Spent Nuclear Fuel and High-Level Radioactive Waste at Yucca Mountain, Nye County, Nevada," Office of Civilian Radioactive Waste Management, vol. II, Appendix J, DOE/EIS-0250D, July 1999.

(DOE, 2002a) "Transporting Spent Nuclear Fuel and High-Level Radioactive Waste to a National Repository: Answers to Frequently Asked Questions," U.S. Department of Energy, Office of Civilian Radioactive Waste Management, Yucca Mountain Project, December 2002 [Available as PDF file through: http://www.ocrwm.doe.gov/wat/transportation.shtml]

(DOE, 2002b) "Final Environmental Impact Statement for a Geologic Repository for the Disposal of Spent Nuclear Fuel and High-Level Radioactive Waste at Yucca Mountain, Nye County, Nevada: Readers Guide and Summary," U.S. Department of Energy, Office of Civilian Radioactive Waste Management, DOE/EIS-0250F, February 2002.

(Federal Register, 1998) "Office of Civilian Radioactive Waste Management; Safe Routine Transportation and Emergency Response Training; Technical Assistance and Funding," vol. 63, No. 83, 30 April 1998.

(FEMA, 1988) "Guidance for Developing State and Local Radiological Emergency Response Plans and Preparedness for Transportation Accidents," Federal Emergency Management Agency, FEMA-REP-5, Draft Revision, August 1988.

(Flynn, 1998) Flynn, James H., C.K. Mertz, and Paul Slovic. "Results of a 1997 National Nuclear Waste Transportation Survey", Decision Research, January 1998 [http://www.state.nv.us/nucwaste/trans/1dr01a.htm]

(FMCSA, 2005) "About Us," Department of Transportation, Federal Motor Carrier Safety Administration website [http://www.fmcsa.dot.gov/aboutus/aboutus.htm]

(FRA, 1998) "Safety Compliance Oversight Plan for Rain Transportation of High-Level Radioactive Waste & Spent Nuclear Fuel," Department of Transportation, Federal Railroad Administration, June 1998, [http://www.fra.dot.gov/downloads/safety/scopfnl.pdf]

(Freking, 1995) Freking, Kevin. "Nuclear plant neighbors worried about waste," *Arkansas Democrat Gazette*, 30 April 1995, p. 1A, col. 1.

(FRERP, 1996) "Federal Radiological Emergency Response Plan," May 1, 1996. Summary available in the *Federal Register*, vol. 61, No. 90, Wednesday, 8 May 1996, Notice: 20 944.

(GAO, 2003) "Spent Nuclear Fuel: Options Exist to Further Enhance Security," General Accounting Office, Report to the Chairman, Subcommittee on Energy and Air Quality, Committee on Energy and Commerce, U.S. House of Representatives, GAO-03-426, July 2003 [http://www.gao.gov/new.items/d03426.pdf]

(Hall, 2002) Hall, Jim. "Testimony of Jim Hall on Behalf of the Transportation Safety Coalition," U.S. Senate, Committee on Energy and Natural Resources, 23 May 2002 [http://www.yuccamountain.org/pdf/hall052302.pdf]

(Halstead, n.d.) Halstead, Robert J. "Radiation Exposures From Spent Nuclear Fuel and High-Level Nuclear Waste Transportation to a Geologic Repository or Interim Storage Facility in Nevada," State of Nevada, Nuclear Waste Project Office, no date [http://www.state.nv.us/nucwaste/trans/radexp.htm]

(Hebert, 2002) Hebert, H. Josef. "Nuclear waste could be terror target," *The Courier*, 20 August 2002, p. 6, col. 1.

(Hebert, 2003) Hebert, H. Josef. "Senate backs plan to revive nuclear power," *The Courier*, 11 June 2003, p. 9, col. 1.

(Holt, 1998) Holt, Mark. "Transportation of Spent Nuclear Fuel," Congressional Research Service, CRS Report for Congress, 97-403 ENR (Updated 29 May 1998) [http://cnie.org/NLE/CRSreports/energy/eng-34.cfm]

(Jordan, 2002) "S.C. to block plutonium trucks," *Arkansas Democrat Gazette*, 15 June 2002, p. 2A, col. 5.

(Lamb, 2001) Lamb, Matthew, Marvin Resnikoff and Richard Moore. "Worst Case Credible Nuclear Transportation Accidents: Analysis for Urban and Rural Nevada," Radioactive Waste Management Associates, August 2001 [http://www.state.nv.us/nucwaste/trans/rwma0108.pdf]

(Macfarlane, 2001) Macfarlane, Allison. "Interim Storage of Spent Fuel in the United States," *Annual Review of Energy and the Environment*, vol. 26, 2001: 201-35.

(Meserve, 2002a) Meserve, Richard A. Letter to Congressman Edward J. Markey, 4 March 2002 [http://www.house.gov/markey/ISS_nuclear_ltr020325a.pdf]

(Meserve, 2002b) Meserve, Richard A. "Nuclear Power Plant Security," Statement submitted by the NRC to the Committee on Environment and Public Works, U.S. Senate, 5 June 2002 [http://www.nrc.gov/reading-rm/doc-collections/congress-docs/congress-testimony/2002/ml021570116.pdf]

(NACO, 2002) "Environment, Energy & Land Use," *The American County Platform and Resolutions*, Resolution 3A, Adopted 16 July 2002 [http://www.naco.org/Content/ContentGroups/Legislative_Affairs/Advocacy1/EELU/EELU1/3EELU_02-03.pdf]

(NCSL, 1998) "Radioactive Materials Transportation: Update for the High-Level Radioactive Waste Working Group," National Conference of State Legislatures, Environment, Energy and Transportation Program, September 1998 [http://www.ncsl.org/programs/esnr/radmatup.htm]

(NEI, 1998) *Industry Spent Fuel Storage Handbook*, Nuclear Energy Institute, NEI 98-01, May 1998.

(Nevada, 1996) "Reported Incidents Involving Spent Nuclear Fuel Shipments 1949 to Present," State of Nevada, 6 May 1996 [http://www.state.nv.us/nucwaste/trans/nucinc01.htm]

(NRC, 2001) "Final Environmental Impact Statement for the Construction and Operation of an Independent Spent Fuel Storage Installation on the Reservation of the Skull Valley Band of Goshute Indians and the Related Transportation Facility in Tooele County, Utah," U.S. Nuclear Regulatory Commission, Office of Nuclear Material Safety and Safeguards, NUREG-1714, vol. 1, December 2001 [http://www.nrc.gov/reading-rm/doc-collections/nuregs/staff/sr1714/v2/sr1714v2.pdf]

(NRC, 2003a) "Safety and Spent Fuel Transportation," U.S. Nuclear Regulatory Commission, NUREG/BR-0292, March 2003 [http://www.nrc.gov/reading-rm/doc-collections/nuregs/brochures/br0292]

(NRC, 2003b) "NRC Increases Secondary Premium for Nuclear Power Plant Insurance," *NRC News*, U.S. Nuclear Regulatory Commission, 4 August 2003 [http://www.nrc.gov/reading-rm/doc-collections/news/2003/03-100.html]

(NRC, 2003c) "Nuclear Security–Before and After September 11," U.S. Nuclear Regulatory Commission [http://www.nrc.gov/what-we-do/safeguards/response-911.html]

(NRC, 2005) "Transporting Spent Nuclear Fuel," Diagram and Description of Typical Spent Fuel Transportation Cask, U.S. Nuclear Regulatory Commission [http://www.nrc.gov/waste/spent-fuel-storage/diagram-typical-trans-cask-system.doc]

(NSC, 2001) "Backgrounder on the Waste Isolation Pilot Plant: How Are Transportation Routes Chosen?" National Safety Council, Environmental Health Center, August 2001 [http://www.nsc.org/public/ehc/rad/backgr8.pdf]

(NTP, 2000) "Questions and Answers About the Program," National Transportation Program, U.S. Department of Energy, November 2000 [http://www.ntp.doe.gov/qa.html]

(NTP, 2005) "National Transportation Program Products & Services," National Transportation Program, U.S. Department of Energy [http://www.ntp.doe.gov/mission.html]

(NWPA, 1982) "Nuclear Waste Policy Act (P.L. 97-425)," as amended, [http://www.ocrwm.doe.gov/wat/index.shtml]

(OCRWM, 2002a) "Final Environmental Impact Statement for a Geologic Repository for the Disposal of Spent Nuclear Fuel and High-Level Radioactive Waste at Yucca Mountain, Nye County, Nevada," Office of Civilian Radioactive Waste Management, U.S. Department of Energy, DOE/EIS-0250, February 2002 [http://www.ocrwm.doe.gov/documents/feis_2/vol_1/ch_06/index1_6.htm]

(OCRWM, 2002b) "Transporting Spent Nuclear Fuel and High-level Radioactive Waste to a National Repository: Answers to Frequently Asked Questions," Office of Civilian Radioactive Waste Management, U.S. Department of Energy [http://www.ocrwm.doe.gov/wat/pdf/snf_transfaqs.pdf]

(OCRWM, 2002c) "Spent Nuclear Fuel Transportation," Office of Civilian Radioactive Waste Management, U.S. Department of Energy [http://www.ocrwm.doe.gov/wat/pdf/snf_trans.pdf]

(OCRWM, 2003) "Cooperative Agreements," Office of Civilian Radioactive Waste Management, U.S. Department of Energy [http://ocrwm.doe.gov/wat/cooperative_agreements.shtml]

(OCRWM, 2005) "Department of Energy Policy Statement for Use of Dedicated Trains for Waste Shipments to Yucca Mountain," Office of Civilian Radioactive Waste Management, U.S. Department of Energy, 2005 [http://www.ocrwm.doe.gov/transport/pdf/dts_policy.pdf]

(OCRWM, 2006) "Transportation of Spent Nuclear Fuel and High-Level Radioactive Waste to Yucca Mountain: Frequently Asked Questions," Office of Civilian Radioactive Waste Management, U.S. Department of Energy, DOE/ONT-0614, Rev. 1, January 2006 [http://www.ocrwm.doe.gov/transport/pdf/snf_transfaqs.pdf]

(Pasternak, 2001) Pasternak, Douglas. "A nuclear nightmare," *U.S. News and World Report*, 17 September 2001, p.44.

(Pasternak, 2003) Pasternak, Douglas. "Assessing threats," *U.S. News and World Report*, 28 April 2003. p. 47.

(PFS, 2003) "FAQs: Transportation," Private Fuel Storage, LLC [http://www.privatefuelstorage.com/faqs/faq/faq-transportation.html]

(PHMSA, 2005a) "About PHMSA," Pipeline and Hazardous Materials Safety Administration, Department of Transportation [http://www.phmsa.dot.gov/about/index.html]

(PHMSA, 2005b) "Hazardous Materials Safety," Office of Hazardous Materials Safety, Pipeline and Hazardous Materials Safety Administration, Department of Transportation [http://www.phmsa.dot.gov/riskmgmt/risk.htm]

(POGO, 2002) "Nuclear Power Plant Security: Voices from Inside the Fences," September 12, 2002, Project on Government Oversight [http://www.POGO.org/p/environment/eo-020901-nukepower.html]

(Public Citizen, 2002) "A Brief History of Irradiated Nuclear Fuel Shipments: Atomic Waste Transport 'Incidents' and Accidents the Nuclear Power Industry Doesn't Want You to Know About," Nuclear Information and Resource Service [http://www.citizen.org/print_article.cfm?ID= 5873]

(Reed, 2000) Reed, James B. "The State Role in Spent Fuel Transportation Safety: Year 2000 Update," NCSL Transportation Series, January 2000, No. 14 [http://www.ncsl.org/programs/esnr/transfer14.pdf]

(Sprung, 2001) Sprung, J.L., D.J. Ammerman, J.A. Koski, and R.F. Weiner. "Spent Nuclear Fuel Transportation Package Performance Study Issues Report," Sandia National Laboratories, NUREG/CR-6768, SAND2001-0821P, 31 January 2001 [http://www.nrc.gov/reading-rm/doc-collections/nuregs/contract/cr6768/nureg-6768.pdf]

(Struglinski, 2005) Struglinski, Suzanne. "NRC: Casks would survive big blaze," *Las Vegas SUN*, 15 September 2005 [http://www.lasvegassun.com/sunbin/stories/sun/2005/sep/15/519360447.html]

(TEC/WG, 2000) "Transportation External Coordination Working Group (TEC) 25-27 July 2000 Meeting Summary," Transportation External Coordination Group, 26 July 2000 [http://www.ntp.doe.gov/tec/ind2000.pdf]

(TEPP, 2002) "Transportation Emergency Preparedness Program," U.S. Department of Energy, Office of Environmental Management, Office of Transportation, February 2002 [http://www.em.doe.gov/otem/TEPPfactsheet02-20-02.pdf]

(Tetreault, 2002) Tetreault, Steve. "NUCLEAR WASTE SHIPMENTS: Report: Department not fully prepared," *Las Vegas Review-Journal*, 9 February 2002 [http://www.energy-net.org/IS/EN/NUKE/WST/FED/NEWS/NP-099.202]

(Travers, 2003) Travers, William D. (2003) "Fitness-for-Duty Enhancements to Address Concerns Regarding Fatigue of Nuclear Facility Security Force Personnel–Draft Order and Compensatory Measures for Power Reactor Licensees," Executive Director for Operations, Nuclear Regulatory Commission, 10 March 2003 (Memorandum COMSECY-03-0012, Attachment 3).

(T-REX, 2000) "A Consideration of Routes and Modes in the Transportation of Radioactive Materials and Waste: An Annotated Bibliography," Transportation Resource Exchange Center, ATR Institute, University of New Mexico, 31 December 2000 [http://www.trex-center.org/dox/routesbib.pdf]

(T-REX, 2003a) "Transportation Resource Exchange Center," Transportation Resource Exchange Center (T-REX) [http://www.trex-center.org]

(T-REX, 2003b) "Spent Nuclear Fuel Transportation Cask Tests," Transportation Resource Exchange Center (T-REX) [http://www.trex-center.org/casktest.asp]

(T-REX, 2003c) "Spent Nuclear Fuel Transportation Regulations," Transportation Resource Exchange Center (T-REX) [http://www.trex-center.org/caskregs.asp]

(USCM, 2002) "Resolution: Transportation of High-Level Nuclear Waste," U.S. Conference of Mayors, Adopted at the 70th Annual Conference of Mayors, Madison, WI, 14-18 June 2002.

(U.S. Congress, 1997) S. 104, Senate, "Nuclear Waste Policy Act of 1997," 105th Congress, 1st Session, [http://thomas.loc.gov/cgi-bin/bdquery/z?d105:SN00104:@@@L&summ2=m&]

(U.S. Congress, 2000) S. 1287, Senate, "Nuclear Waste Policy Amendments Act of 2000," 106th Congress, 1st Session [http://thomas.loc.gov/cgi-bin/bdquerytr/z?d106:SN01287:@@@D&summ2=m&]

(U.S. Congress, 2003) S. 14, Senate, "A Bill to Enhance Energy Security of the United States," 108th Congress, 2nd Session [http://thomas.loc.gov/cgi-bin/bdquery/z?d108:s.00014:]

(WGA, 2001) "Assessing the Risks of Terrorism and Sabotage Against High-Level Nuclear Waste Shipments to A Geologic Repository or Interim Storage Facility," Policy Resolution 01-03, 14 August 2001 [http://www.westgov.org/wga/policy/01/01_03.pdf]

(WIEB, 1996) "HLW Committee Comments on DOE's Notice of Proposed Policy and Procedures for Safe Transportation and Emergency Response Training for Spent Nuclear Fuel and High-Level Radioactive Waste (Notice)," Western Interstate Energy Board, 12 September 1996 [http://www.westgov.org/wieb/reports/180c1998.htm]

(WIEB, 1998) "HLW Committee Comments on DOE's April 30, 1998 Notice of Revised Policy and Procedures for Safe Transportation and Emergency Response Training; Technical Assistance and Funding for Spent Nuclear Fuel and High-Level Radioactive Waste (Notice)," Western Interstate Energy Board, 31 July 1998 [http://www.westgov.org/wieb/reports/180c1998.htm]

(WIEB, 2002) "Reports and Comments," Western Interstate Energy Board, website [http://www.westgov.org/wieb/reports.html]

(Witt, 2002) Witt, James Lee. *Stronger in the Broken Places* (Times Books, NY: 2002).

(Witt, 2003) Witt, James Lee. "Review of Emergency Preparedness of Areas Adjacent to Indian Point and Millstone," James Lee Witt Associates, LLC, 7 March 2003 [Available at: http://wittassociates.com/news_final_report_NY.html]

(Zacha, 2003) Zacha, Nancy J. "Fear of Shipping," Editor's Note, *Radwaste Solutions*, March/April 2003 [http://www2.ans.org/pubs/magazines/rs/pdfs/2003-3-4-2.pdf]

Chapter 6

The Conceptual Framework

The problem of continued highly radioactive waste generation, coupled with the inability to find a long-term solution for the problem, has both homeland security and energy security overtones. At the same time, the political conflict with each storage or disposal option has prevented developing a long-term solution to the problem.

This chapter presents a theoretical framework for assessing homeland security and energy security concerns, while at the same time factoring in the political considerations associated with the storage and disposal of spent nuclear fuel and high-level waste. Basically, there are three options in dealing with highly radioactive waste:[1]

- continue the current on-site storage regime
- find an acceptable centralized interim storage solution
- achieve an acceptable permanent disposal solution

While homeland security and energy security concerns should determine highly radioactive waste policy, in reality, it is the political considerations that have been the driving force behind policy decisions.

Determining Risk

Determining risk can be a difficult undertaking due to the number of variables that can exist simultaneously. For example, for obvious operational security reasons, it is difficult to obtain information on the security practices and the overall effectives of security at commercial nuclear power plants and defense facilities. Thus, it is difficult to develop an accurate understanding of the actual homeland security risks at nuclear facilities. This, in turn, complicates the analytical process. Nevertheless, subjective evaluations of potential security and political risks associated with highly radioactive waste storage/disposal options can be made. And, ultimately, employing risk assessment can be used to assist policymakers in choosing the best policy option for coping with highly radioactive waste.

Risk assessment has gained widespread use in both the government and private sectors as a means to maximize benefits and minimize consequences. Today, a number of U.S. government agencies are now employing a variety of risk analysis

---

1   While processing could be considered a fourth option, the U.S. abandoned reprocessing in the 1970s due primarily to concerns over proliferation. However, other countries such as France, Russia, and Japan view spent nuclear fuel as a resource to be reprocessed, which ultimately reduces, but does not eliminate, the amount of highly radioactive waste.

strategies for coping with radioactive waste.[2] For example, the Department of Energy (DOE) has commissioned a study to develop technically sound "risk-based" approaches for the disposition of transuranic and high-level radioactive waste.[3] The Environmental Protection Agency (EPA) uses the "Superfund Risk Assessment Model" to determine the ecological and health risks posed by a particular hazardous waste site (e.g., nuclear waste).[4] The Department of Homeland Security uses a variety of models and simulations to evaluate risks associated with particular critical infrastructure vulnerabilities (e.g., nuclear power plants) in order to make more informed protection decisions.[5] The Nuclear Regulatory Commission (NRC) began to focus more on probabilistic risk assessment in the 1990s as the NRC moved toward "risk-informed regulation" related to nuclear safety issues.[6]

For the purposes of this framework, *risk* is defined as the potential for a possible event to occur, coupled with the possible consequence of that event. Risk analysis can use either a quantitative approach (e.g., probabilistic modeling such as Monte Carlo analysis) or a qualitative approach (e.g., intituive analysis). A qualitative approach is used in this chapter by subjectively assessing the potential homeland security and energy security risks associated with the current highly radioactive waste storage and disposal options (on-site storage, centralized interim storage, or permanent disposal). In addition, the political risks associated with each storage and disposal option are evaluated. Clearly, homeland security and energy security concerns should determine policy. But, in reality, political considerations have had a major impact on policy concerning highly radioactive waste.

---

2   See (NAS, 1994) "Building Consensus Through Risk Assessment and Management of the Department of Energy's Environmental Remediation Program," Committee to Review Risk Management in the DOE's Environmental Remediation Program, National Research Council, National Academy Press, 1994, for a discussion on using risk assessment for cleaning up DOE sites. [http://www.nap.edu/openbook/NI000191/html/R1.html]

3   See (NAS, 2005a) "'Risk-Based' Approaches for Disposition of Transuranic and High-Level Radioactive Waste," Current Projects, BRWM-U-03-01-A [http://www.nas.edu] for a synopsis of the project. The published report (*Risk and Decisions About Disposition of Transuranic and High-Level Radioactive Waste*) is available at [http://newton.nap.edu/catalog/11223.html#toc].

4   See (EPA, 2005) "Superfund Risk Assessment," U.S. Environmental Protection Agency, Waste and Cleanup Risk Assessment [http://www.epa.gov/oswer/riskassessment/risk_superfund.htm]

5   See (DHS, 2003) "The National Strategy for the Physical Protection of Critical Infrastructure and Key Assets," February 2003, [http://www.dhs.gov/xlibrary/assets/Physical_Strategy.pdf].

6   See Jackson, Shirley M. (1998), "Transitioning to Risk-Informed Regulation: The Role of Research," U.S. Nuclear Regulatory Commission, Office of Public Affairs, 26th Annual Water Reactor Safety Meeting, 26 October 1998, No. S-98-26 [http://www.nrc.gov/reading-rm/doc-collections/commission/speeches/1998/s98-26.html]. Also see, Callan, Joseph L. (1998), "Risk-Informed, Performance-Based and Risk-Informed, Less-Prescriptive Regulation in the Office of Nuclear Material Safety and Safeguards," U.S. Nuclear Regulatory Commission, No. SECY-98-138, 11 June 1998 [http://www.nrc.gov/reading-rm/doc-collections/commission/secys/1998/secy1998-138/1998-138scy.html]

The "security risk" assessment models use a qualitative analysis of the homeland security and energy security concerns associated with a particular radioactive waste storage or disposal option. The following definitions are being used in the security risk assessment models (*poor, fair, good* and *excellent*):

- *poor security risk* means that the storage/disposal option does not adequately address potential homeland security or energy security risks
- *fair security risk* partially addresses the security risk under consideration (e.g., may only be a partial or temporary fix)
- *good security risk* means that the security risk is generally addressed but with some reservations (e.g., there still may be some homeland or energy security concerns that will occur in the long-term)
- *excellent security risk* addresses the security risk for the long-term.

These risk assessment models have been devised for their simplicity and fit with the homeland security and energy security concerns associated with the on-site storage, centralized interim storage, and permanent disposal methods of coping with highly radioactive waste.

The "political risk" assessment model also uses a qualitative analysis of the political conflict generated by a particular radioactive waste storage/disposal policy decision. The following definitions are being used in the political risk analysis model (*low, moderate* and *high*):

- *low political risk* involves minimum conflict (e.g., only limited, generally localized conflict)
- *moderate political risk* involves mid-level conflict (e.g., conflict is more evident, usually at the state level)
- *high political risk* involves significant political conflict (e.g. substantial conflict existing at the national level).

This political risk model also has been devised for its simplicity and fit with level of political conflict associated with the on-site storage, centralized interim storage, and permanent disposal options for highly radioactive waste.

Homeland Security Concerns

While security has been intensified since 9/11, some facilities have more robust security than others. And clearly, storing large quantities of highly radioactive waste at numerous locations across the country heightens anxiety over a possible terrorist attack with the potential for a significant release of radiation. In 2002, Secretary of Energy Spencer Abraham cited concerns over maintaining large amounts of highly radioactive waste at over 130 sites in 39 states within 75 miles of over 160 million Americans as a threat to homeland security. (Abraham, 2002a: 3) Thus, it would seem to be prudent to minimize the number of potential targets by removing as much of the spent nuclear fuel and high-level waste as possible.

Over the years, numerous threats have been directed at commercial nuclear power facilities. Moreover, concern over security has increased since the terrorist attacks on 9/11 – especially since nuclear power facilities have been singled out as a prime target for a future attack. According to a former Department of Energy official, "These are the ultimate dirty bombs. Let's not pretend the way we are storing this waste is safe and secure in an age of terrorism." (Vartabedian, 2005: 1)

Compounding this concern, in 2002, the Director of the U.S. Nuclear Regulatory Commission (NRC) acknowledged in a report to Congress that security "weaknesses" were identified in almost half of the mock security exercises conducted at nuclear power plants during the 1990s. (Meserve, 2002a: 27) At the same time, the watchdog group Project on Government Oversight (POGO) asserted that post-9/11 interviews with security guards at several commercial nuclear power facilities revealed that they were under-manned, under-trained, under-equipped, and unsure how to respond to incidents requiring deadly force. (POGO, 2002: 2-3) A 2004 Government Accountability Office (GAO) review of NRC efforts to improve post-9/11 security at commercial nuclear power facilities noted that while the NRC responded "quickly and decisively" after 9/11, the NRC still was not able to independently determine "that each plant has taken reasonable and appropriate steps" to protect against the perceived threats. (GAO, 2004: i) While some of this concern over security has been directed towards the reactors themselves, they are less of a problem since they are housed within steel and concrete reinforced containment buildings. However, the highly radioactive waste stored on-site is more vulnerable since the spent nuclear fuel is stored either in cooling pools within non-reinforced buildings or stored in dry casks above ground. Thus, the spent fuel is more worrisome since it would be more vulnerable to any terrorist attack.

Not everyone believes that the radioactive waste stored on-site at nuclear power facilities is a primary security risk. Nuclear industry executives and government officials refute assertions that the waste stored on-site is a prime security worry. For example, according to a senior Department of Homeland Security official, the on-site highly radioactive waste is not considered to be a high-value target. The view is that increased security since 9/11, coupled with the belief that on-site storage methods are sufficiently protected to prevent a 'catastrophic' release of radiation, would deter terrorists from attacking a nuclear facility. (Stanton, 2005) Moreover, the "weaknesses" identified during mock security exercises could point to the fact that the exercises are designed to test the limits of security preparedness. According to the government official, it demonstrates the validity of the exercises. (Stanton, 2005)

Clearly, steps have been taken recently to improve security at commercial nuclear power facilities. For example, according to the Nuclear Energy Institute (NEI) – the nuclear power industry's policy organization – over $370 million has been invested since 9/11 in enhancing security at nuclear power plants, and the commercial nuclear power industry has increased security forces by about one-third (8,000 officers). (NEI, 2004b: 1) At the same time, the federal government has taken steps to improve security such as establishing an Office of Nuclear Security and Incident Response to consolidate NRC safeguards and incident response. (Meserve, 2002b: 2-3) In 2003, the NRC issued a new set of *design based threat (DBT) guidelines* requiring commercial nuclear power plants to implement new security plans by the end of 2004 that are capable of defending against potential threats to security. The new DBT guidelines reflect "the increased size of a potential terrorist force, the more sophisticated weaponry, and the different methods of deployment demonstrated by the September 11 terrorist attacks." (GAO, 2004: 7)

On-Site Storage

In spite of the attempts to improve security, there still is a number of specific homeland security concerns associated with large quantities of highly radioactive waste stored on-site. In fact, a 2005 National Academy of Sciences report stated that spent fuel storage facilities are possible terrorist targets because of the potential for a release of radiation. (NAS, 2005b: 4) First, there is the potential for a significant release of radiation if the highly radioactive waste on-site was attacked or sabotaged by a terrorist group. For example, if terrorists were able to breach a cooling pool filled with highly radioactive spent fuel rods, there would be the potential for a significant release of radiation due to a loss of water to keep the intensely hot fuel rods sufficiently cooled. Moreover, the dry cask storage method also presents some risk. While the dry casks used to store spent nuclear fuel are robust and designed to take considerable punishment, concern has been voiced over whether they would be able to withstand a terrorist assault such as an air attack using fully fueled jet airliners.[7] This could be especially serious if the attack was near a major metropolitan center. Second, although small, there is the potential for the theft and subsequent use – most likely as a "dirty bomb" – of the nuclear material by terrorists. While the highly radioactive waste would be difficult to handle due to its lethality and sheer volume and weight, individuals willing to sacrifice themselves for a cause could pose a possible threat. Third, there is the potential for nuclear proliferation if stolen highly radioactive waste was transferred to a rogue state or terrorist group. As a result, the overall homeland security risk of the current on-site storage regime is assessed as poor since this option does not adequately address the security problem.

Interim Storage

The construction of a centralized interim storage facility for spent nuclear fuel would address, at least for the mid-term, some of the homeland security concerns associated with the current on-site storage regime by reducing the number of sites (i.e., the signature) vulnerable to a terrorist attack, especially for those facilities located near large metropolitan centers. Moreover, security at a centralized facility likely would be more robust given the fact that significant quantities of highly radioactive waste would be stored there and the facility would be considered a high-value target. However, since a centralized facility would not have the capability to store all of the on-site highly radioactive waste, there would continue to be some spent nuclear fuel

---

7 According to a senior homeland security official, however, models involving attack scenarios by various types of aircraft did not result in any 'catastrophic' release of radiation. (Stanton, 2005)

and high-level waste stored on-site in dry casks at a number of reactor sites across the country.[8] Moreover, any centralized option would require the transport of highly radioactive waste across great distances, sometimes near heavily populated centers.[9] While DOE characterizes the nuclear transportation record as "impressive," there still is the potential for an accident or terrorist attack. (DOE, 2002: 2) Finally, any centralized interim storage option would be a potential terrorist target due to the large quantity of highly radioactive waste stored above ground. Thus, while the initial homeland security assessment for a centralized interim storage facility would appear to be good, the continued requirement for some on-site dry cask storage, the need to transport significant quantities of spent nuclear fuel, and the potential security risk of such a facility would lower the overall homeland security risk assessment to fair since this option would only address the problem partially.

Permanent Disposal

Permanent disposal provides the greatest protection for homeland security. Since spent nuclear fuel and high-level waste would be stored below ground in a highly secured facility, there is little risk of a terrorist attack, sabotage, or theft of the nuclear material. Moreover, the most vulnerable waste (dry casks) could be removed from all of the reactor sites. In addition, there are national security concerns that a permanent repository would be able to address. For example, a permanent repository would guarantee a secure route for spent nuclear fuel from the Navy's nuclear power fleet. Moreover, a permanent repository also would support non-proliferation objectives by disposing of weapons grade plutonium from decommissioned nuclear weapons. This would preclude the possibility of the diversion of the highly prized nuclear material to unauthorized parties.

However, there will still continue to be spent nuclear fuel stored on-site in cooling pools. Thus, while the homeland security risk associated with permanent disposal initially would appear to be excellent, there still would be some highly radioactive waste stored on-site near the reactors, and the transport of the highly radioactive waste over long distances though multiple jurisdictions would continue to pose some safety and security concerns. Therefore, similar to transportation concerns associated with a centralized interim storage option, the overall assessment of a permanent disposal option has to be somewhat tempered. As a result, the overall homeland security risk for a national highly radioactive waste repository is assessed only as very good, since it generally addresses the problem. See Figure

---

8   For example, while the two proposals (Skull Valley, UT; and the Owl Creek, WY) being pursued plan to store up to 40,000 metric tons of spent nuclear fuel, it is likely that only the Skull Valley initiative could be successful – at least in the near-term. (PFS, 2000: 1; OECP, 2000: 3) Moreover, legal challenges to the Skull Valley project may reduce its capacity, which would further limit the number of dry casks that could be stored. (PFS, 2003: 2)

9   See (Rogers and Kingsley, 2004b) Rogers, Kenneth A. and Marvin G. Kingsley. "The Transportation of Highly Radioactive Waste: Implications for Homeland Security," Journal of Homeland Security and Emergency Management, vol. 1, Issue 2, Article 13, 2004 [http://www.bepress.com/jhsem/vol1/iss2/13] for a discussion of the risks associated with highly radioactive waste transportation.

6.1 (Homeland Security Risk Analysis) for a graphic summary of the homeland security risk associated with on-site storage, centralized interim storage, and permanent disposal.

	Poor	*Fair*	*Good*	*Excellent*
On-site Storage	X			
Interim Storage		X		
Permanent Disposal			X^*	

Figure 6.1 Homeland Security Risk Analysis
* Denotes that homeland security would be *very good*.

Energy Security Concerns

Concerns over adequate energy supplies first surfaced in the United States during the early 1970s with the Arab oil embargo. Oil shocks in subsequent years and the recent dramatic rise in oil prices have reinforced the perception of the need for the United States to reduce its dependency on foreign sources of energy. However, this can only be done if the United States does not lose any of its current domestic energy generation capacity and is able to keep pace with future energy demands. According to the U.S. Energy Information Administration (EIA), U.S. electricity generation is projected to increase by about 50 percent by 2025. (EIA, 2004: 145) In response to the anticipated demand for new energy sources, the Bush Administration has called for the expansion of nuclear generated electricity as part of its national energy policy.[10] In support of the new energy plan, DOE has developed the Nuclear Power 2010 program, which would facilitate the licensing process for building new nuclear power plants. (DOE, 2004b: 1) At the same time, the commercial nuclear power industry has developed a plan (Vision 2020) to increase nuclear electricity generation by 50,000 megawatts from new nuclear power plant capacity by 2020. (NEI, 2004a: 1) In any case, whether nuclear power generation is increased to meet projected future energy demands, or even if just current production levels are maintained, the

---

10 See (White House, 2001) "National Energy Policy" for a discussion of the Bush Administration energy plan. [http://www.whitehouse.gov/news/releases/2001/05/20010517-2.html] Also, see "DOE Researchers Demonstrate Feasibility of Efficient Hydrogen Production from Nuclear Energy," Press Release, 30 November 2004 [http://www.energy.gov/news/1545.htm] for a discussion of using new generation nuclear reactor technologies to produce hydrogen for powering future vehicle fuel cell technology. (DOE, 2004a: 1)

amount of spent nuclear fuel will continue to accumulate at commercial reactor sites across the country further straining on-site storage capabilities.

On-Site Storage

Currently, spent nuclear fuel is being stored on-site awaiting transfer to either an interim storage facility or a national repository. However, there is only limited on-site storage space available either in the cooling pools or in the dry casks. According to the nuclear power industry, by 2010, 78 (over 75 percent) commercial nuclear power plants will not have any storage capacity left in their cooling pools. (NEI, 2004c: 1) While dry cask storage is now being used at a number of locations to supplement their cooling pool capacity, some states have attempted to limit the development or expansion of dry storage. (NEI, 2004c: 2) Thus, a lack of sufficient on-site storage for the spent nuclear fuel could force the commercial nuclear power industry to shut down some of its reactors prematurely in order not to exceed the limited storage capability. Energy Secretary Abraham expressed concern over the lack of sufficient on-site storage capacity and the potential for a premature shutdown of commercial nuclear reactors:

> A repository is important to our energy security. We must ensure that nuclear power, which provides 20 percent of the nation's electric power, remains an important part of our domestic energy production. Without the stabilizing effects of nuclear power, energy markets will become increasingly more exposed to price spikes and supply uncertainties. (Abraham, 2002a: 2-3)

Abraham noted in subsequent testimony before Congress that nuclear waste continues to pile up at commercial nuclear sites that are in his words, "running out of room for it." (Abraham, 2002b: 3) Abraham's remarks highlight a need to establish a secure pathway for spent nuclear fuel in order to ensure that current commercial nuclear power facilities are not forced to shut down their reactors prematurely. Thus, if U.S. commercial nuclear energy power production were curtailed due to a lack of sufficient on-site storage space then the overall energy security risk of the current on-site storage regime would be assessed as being poor.

Interim Storage

Energy security concerns involved with centralized interim storage of highly radioactive waste would be partially addressed, at least for the mid-term. Providing a temporary pathway for spent nuclear fuel would prevent commercial nuclear power facilities from potentially having to shut down their reactors prematurely due to a lack of sufficient on-site storage space. If the reactor site was in danger of reaching its storage capacity for spent nuclear fuel, then the highly radioactive waste could be transported to the centralized interim storage facility, freeing up additional on-site storage space. However, since a centralized interim storage facility would only be a partial fix for the mid-term and not for the long-term, the overall energy security risk for a centralized interim storage option is assessed only as good.

Permanent Disposal

Energy security would be increased greatly if a national repository were constructed since it would provide a long-term pathway for spent nuclear fuel, preventing the potential for a premature shutdown of commercial nuclear reactors due to a lack of sufficient on-site storage space. Since large quantities of on-site spent nuclear fuel (especially dry cask storage) could be removed, the overall energy security risk for a national highly radioactive waste repository is assessed as excellent. See Figure 6.2 (Energy Security Risk Analysis) for a graphic summary of the energy security risk associated with on-site storage, centralized interim storage, and permanent disposal.

	Poor	*Fair*	*Good*	*Excellent*
On-site Storage	X			
Interim Storage			X	
Permanent Disposal				X

Figure 6.2 Energy Security Risk Analysis

Political Risk

The issue of highly radioactive waste storage and disposal is inherently conflictual for a number of reasons. First, the costs of managing the nuclear waste are tangible (i.e., readily apparent) and substantial, while the benefits of finding a solution, such as enhancing security, are intangible (i.e., not readily apparent). Second, the costs incurred for centralized storage and disposal are apparent immediately, while the benefits will be realized sometime in the future. Third, since costs are concentrated in the state and community that host the centralized interim storage or permanent disposal facility, and the benefits of increased homeland and energy are dispersed to the country as a whole, conflict at the state or local level is almost inevitable. When policymakers perceive issues to be of this nature, they generally prefer to defer decisions in order to minimize conflict. Paradoxically, as the need for action has become more acute, the conflict generated by the politicization of the issue has delayed progress on developing a viable long-term solution to the problem.

On-Site Storage

Conflict associated with the current on-site storage regime has been relatively muted. Thus, continuing a relatively non-controversial status quo policy (especially one that has been in place for a number of years) involves minimal short-term political risk. Policymakers generally prefer to follow existing policy, or making only minor policy adjustments (i.e., incremental policymaking). However, that does not imply that there has not been conflict over the on-site storage of highly radioactive waste.

While there has been little general public opposition to the storing of spent nuclear fuel at commercial nuclear power facilities and DOE research and weapons facilities, there have been some concerns voiced by some citizen groups over the safety and security of the on-site waste. For example, several safety incidents have been documented for some dry cask storage containers such as defective welds, cracked seals, and explosions. (Macfarlane, 2001: 218) Moreover, POGO has released reports critical of on-site security at several commercial nuclear power facilities. In addition, the commercial nuclear power industry has brought legal action against the federal government for its failure to take title to the waste as mandated by the 1982 Nuclear Waste Policy Act (NWPA). At the same time, the Bush Administration's energy plan calls for increasing future nuclear power generation capacity. This would be difficult to achieve if some commercial nuclear energy facilities were forced to shut down reactors due to a lack of sufficient on-site storage space. Thus, while conflict currently is relatively muted, it would be reasonable to expect conflict to increase in the future as on-site storage capacity reaches it limit. In fact, the NRC has pointed out that some opposition to the dry cask storage method has made licensing operations more difficult in a number of states. (NRC, 1996: 7) As a result, while the overall political risk associated with continuing temporary on-site storage currently is assessed to be low, the lack of resolution of this issue ultimately will increase conflict and political risk.

Interim Storage

The concept of centralized interim highly radioactive waste storage has been controversial since opponents have voiced concern that any interim facility could become a de facto repository.[11] Thus, any interim storage solution has been inextricably tied to progress on long-term disposal. While a centralized interim storage solution will address some of the more immediate problems associated with the continued on-site storage regime, it does not permanently solve the problem of what to do with the highly radioactive waste and offers only a temporary solution.

Although a centralized storage option was first proposed in the early 1970s, conflict and politicization of the issue have prevented the construction of any centralized facility. Initial efforts by local jurisdictions to consider hosting a centralized interim storage facility generally were blocked by the host state. Thus, while a centralized storage option for highly radioactive waste would partially address both homeland security and energy security concerns, the politicization of the issue has produced considerable conflict in the states that would possibly host such a facility. As a

---

11 See (Rogers and Kingsley, 2004a) Rogers, Kenneth A. and Marvin G. Kingsley, "The Politics of Interim Radioactive Waste Storage: The United States," *Environmental Politics*, vol. 13, No. 3, September 2004: 590-611, for an analysis of the politics associated with interim radioactive waste storage.

result, the overall political risk associated with constructing a centralized interim storage facility for spent nuclear fuel is assessed to be moderate.

Permanent Disposal

The permanent disposal of highly radioactive waste has been the most politically contentious option primarily because the "not in my backyard" (NIMBY) reaction has generated opposition in states designated as a potential host for a nuclear waste repository. In an effort to address the political opposition that surfaced with respect to disposing of highly radioactive waste, the 1982 Nuclear Waste Policy Act established procedures and timelines for constructing and operating a national repository. In accordance with the 1982 Act, a site was first to be identified in the West with an additional site in the East to be designated later. The site selection process was slow and arduous due to opposition generated by the states that were initially identified as a potential host for the repository.[12] To circumvent this opposition, Congress passed the 1987 Nuclear Waste Policy Amendments Act, which singled out Yucca Mountain, Nevada, as the only site to be studied. Many observers have criticized the site selection process; and even the Director of DOE's Civilian Radioactive Waste Management Program at the time commented that Nevada was "slam dunked." (Suplee, 1995: A18) Clearly, Nevada with a small population and junior congressional delegation at the time did not have the political influence that the other potential host states such as Texas and Washington had. Thus, resentment over the perceived unfairness of the site selection process coupled with apprehension over the geologic suitability of the Yucca Mountain site has generated additional controversy. In response, Nevada has attempted to block the initiative at every turn, which has considerably slowed the pace of the project.[13] As a result, the attempt to construct a geologic repository has become intensely politicized; and thus, the political risk is assessed as high. See Figure 6.3 (Political Risk Analysis) for a graphic summary of the political risk associated with on-site storage, centralized interim storage, and permanent disposal.[14]

---

12 Six western states were identified (all with prior nuclear industry involvement) as a possible host for a national repository: Louisiana, Mississippi, Nevada, Texas, Utah and Washington. Potential host states identified in the East were: Georgia, Maine, Minnesota, New Hampshire, North Carolina, Virginia and Wisconsin. (Kraft, 1996: 114-115)

13 The elevation of Senator Harry Reid (D, NV) to the position of majority leader in the Senate could further slow the Yucca Mountain project. Senator Reid has been quite successful in the past in reducing funding for the repository.

14 See (Rogers and Rogers, 1998) Rogers, Kenneth A. and Donna L. Rogers. "The Politics of Long-Term Highly Radioactive Waste Disposal," *Midsouth Political Science Review*, vol. 2, 1998: 61-71 for an analysis of the politics associated with the Yucca Mountain project.

	Low	*Moderate*	*High*
On-site Storage	X*		
Interim Storage		X	
Permanent Disposal			X

Figure 6.3 Political Risk Analysis for Highly Radioactive Waste
* Denotes that increased political conflict is likely in the future as storage space becomes more limited and likely increasing the political risk of on-site storage.

Summary

The continued on-site storage of highly radioactive waste at numerous locations around the country (some near large metropolitan centers) presents an unacceptable homeland security risk due to the potential for a terrorist attack, sabotage, or theft of the material. However, homeland security concerns could be mitigated somewhat by moving more vulnerable dry casks from above ground to a more protected area on-site until either a centralized interim storage facility or a permanent national repository is ready to accept highly radioactive waste. Even if the most vulnerable waste is moved to a shielded area, the current on-site storage regime still does not adequately address homeland security concerns and thus, is assessed as poor. In addition, the current on-site storage regime provides poor energy security due to the potential for having to shut down some commercial nuclear energy reactors at facilities where there is a lack of adequate on-site spent fuel storage space. The possible premature closing of commercial nuclear power facilities due to a shortage of adequate on-site spent nuclear fuel storage poses an unacceptable energy security risk during a time when the U.S. cannot afford to increase its dependency on foreign sources of energy.

Centralized interim storage would address some homeland security concerns. While some spent nuclear fuel would remain on-site, the most vulnerable on-site waste could be moved to a more secure centralized facility. However, significant quantities of spent nuclear fuel would have to be transported to any centralized interim storage facility. Moreover, a centralized facility with a large quantity of aboveground storage would be an attractive terrorist target. Thus, security concerns associated with continued on-site dry cask storage as well as transportation concerns would lower the overall homeland security risk to fair. Energy security at a centralized interim storage facility would rate better since an interim storage facility would provide a temporary pathway for spent nuclear fuel, which would postpone the shutdown of reactors prematurely due to a lack of sufficient on-site storage space. However, since a centralized interim storage facility would only result in a partial fix for the mid-term (not the long-term) the overall energy security assessment for a centralized interim storage option only would be considered good.

A national repository for the permanent disposal of highly radioactive waste (e.g., Yucca Mountain) would provide excellent homeland security since the nuclear waste would be stored below ground in a highly secured facility where the risk of sabotage or a terrorist attack would be minimal. However, a national repository would require transporting large quantities of highly radioactive waste across the country. Thus, while a secured national repository would provide excellent protection against a terrorist attack, the overall homeland security risk is assessed only as very good due to transportation security concerns. A national repository would provide a long-term pathway for spent nuclear fuel alleviating the need to shut down commercial nuclear power reactors prematurely due to a lack of sufficient on-site nuclear waste storage space. Thus, overall energy security would be assessed as excellent.

While the homeland security and energy security concerns should drive policy making, the reality is that political considerations have been a major determinant of highly radioactive waste storage/disposal policy. Opposition to constructing both a centralized interim storage facility and a permanent disposal facility has led to the policy of storing the waste on-site. The current on-site storage regime involves the least amount of conflict and is most acceptable politically since it follows a status quo policy; and thus, the political risk is assessed to be low. However, it is likely that conflict and the associated political risks will increase as adequate on-site storage space becomes more problematical. On the other hand, the centralized interim storage option has been controversial since opponents have voiced concern over the safety of the facility, as well as the fact that the interim facility could become a de facto repository if a national repository is not completed as planned. Moreover, opposition is intensified by the fact that any centralized option would require transporting significant amounts of highly radioactive waste to the facility. In addition, since an interim solution does not conclusively solve the problem of what to do with the highly radioactive waste, opponents argue that policymakers should keep nuclear waste on-site until a permanent option is found. As a result, the political risk is assessed to be moderate. The search for a geologic repository has been even more contentious and politicized. Opposition sparked by the perceived unfairness of the site selection process and scientific uncertainty over the geologic suitability of the Yucca Mountain facility has intensified the political conflict. As a result, the political risk is assessed to be high. See Figure 6.4 (Overall Homeland/ Energy Security – Political Risk Analysis for Highly Radioactive Waste) for a summary of the overall risk analysis.

Thus, it becomes apparent that while security considerations would dictate that a centralized storage/disposal option should be made available in the near term, political realities have prevented this from happening. More importantly, these same political realities will clearly be an obstacle to developing a long-term solution to mitigating both homeland security and energy security concerns.

The inability of U.S. policymakers to develop an interim storage solution or a permanent disposal option has important homeland security and energy security implications. Chapter 7 gives a brief summary of the issue of highly radioactive waste storage and disposal, reviews the factors that affect U.S. radioactive waste policy, explores the alternatives that can be taken to address the problem of the mounting inventories of highly radioactive waste, and makes recommendations that would help to reduce the impact of the delay in achieving a coherent long-term policy on what to do with the lethal waste.

	Homeland Security & Political Risk	Energy Security & Political Risk
On-site Storage	*Poor Homeland Security Risk* *&* *Low Political Risk**	*Poor Energy Security Risk* *&* *Low Political Risk**
Interim Storage	*Fair Homeland Security Risk* *&* *Moderate Political Risk*	*Good Energy Security Risk* *&* *Moderate Political Risk*
Permanent Disposal	*Very Good Homeland Security Risk* *&* *High Political Risk*	*Excellent Energy Security Risk* *&* *High Political Risk*

Figure 6.4 Interpolated Homeland/Energy Security and Political Risk Analysis for Highly Radioactive Waste
* Denotes increased political conflict likely in the future.

Key Terms

risk
poor security risk
fair security risk
good security risk
excellent security risk
low political risk
moderate political risk
high political risk
design based threat guidelines

Useful Web Sources

U.S. Environmental Protection Agency [http://www.epa.gov]
U.S. Department of Energy [http://www.doe.gov]
U.S. Department of Energy, Energy Information Administration [http://www.eia.doe.gov]
U.S. Department of Homeland Security [http://www.dhs.gov]
U.S. Nuclear Regulatory Commission [http://www.nrc.gov]
National Academy of Sciences [http://www.nas.edu]
Nuclear Energy Institute [http://www.nei.org]
Private Fuel Storage [http://www.privatefuelstorage.com]

References

(Abraham, 2002a) Abraham, Spencer. Letter to President Bush recommending approval of Yucca Mountain for the development of a nuclear waste repository, Secretary of Energy, U.S. Department of Energy, 14 February 2002.

(Abraham, 2002b) Abraham, Spencer. "A Review of the President's Recommendation to Develop a Nuclear Waste Repository at Yucca Mountain, Nevada," Secretary of Energy, Prepared Witness Testimony, Committee on Energy and Commerce, Subcommittee on Energy and Air Quality, 18 April 2002 [http://energycommerce.house.gov/107/hearings/04182002Hearing505/Abraham859print.htm]

(Callan, 1998) Callan, Joseph L. "Risk-Informed, Performance-Based and Risk-Informed, Less-Prescriptive Regulation in the Office of Nuclear Material Safety and Safeguards," U.S. Nuclear Regulatory Commission, No. SECY-98-138, 11 June 1998 [http://www.nrc.gov/reading-rm/doc-collections/commission/secys/1998/secy1998-138/1998-138scy.html]

(DHS, 2003) "The National Strategy for the Physical Protection of Critical Infrastructure and Key Assets," U.S. Department of Homeland Security, 14 February 2003, [http://www.dhs.gov/xlibrary/assets/Physical_Strategy.pdf]

(DOE, 2002) "Spent Nuclear Fuel Transportation," U.S. Department of Energy, Office of Public Affairs [http://www.ocrwm.doe.gov/transport/pdf/snf_trans.pdf]

(DOE, 2004a) "DOE Researchers Demonstrate Feasibility of Efficient Hydrogen Production from Nuclear Energy," Press Release, 30 November 2004 [http://www.energy.gov/news/1545.htm]

(DOE, 2004b) "Nuclear Power 2010," Department of Energy, Office of Nuclear Energy, Science and Technology, updated: September 2006 [http://np2010.doc.html]

(EIA, 2004) "Annual Energy Outlook 2004," Energy Information Administration, Table A8 (Electric Supply, Disposition, Prices and Emissions): 145 [http://www.eia.doe.gov/oiaf/archive/aeo04/pdf/0383(2004).pdf]

(EPA, 2005) "Superfund Risk Assessment," U.S. Environmental Protection Agency, Waste and Cleanup Risk Assessment [http://www.epa.gov/oswer/riskassessment/risk_superfund.htm]

(GAO, 2004) "Nuclear Regulatory Commission: Preliminary Observations on Efforts to Improve Security at Nuclear Power Plants," Government Accountability Office, Statement of Jim Wells (Director, National Resources and Environment), Testimony Before the Subcommittee on National Security, Emerging Threats, and International Relations, Committee on Government Reform, House of Representatives, GAO-04-106T, 14 September 2004 [http://www.gao.gov/new.items/do041064t.pdf]

(Jackson, 1998) Jackson, Shirley M. "Transitioning to Risk-Informed Regulation: The Role of Research," U.S. Nuclear Regulatory Commission, Office of Public Affairs, 26th Annual Water Reactor Safety Meeting, 26 October 1998, No. S-98-26 [http://www.nrc.gov/reading-rm/doc-collections/commission/speeches/1998/s98-26.html]

(Kraft, 1996) Kraft, Michael E. "Democratic Dialogue and Acceptable Risks: The Politics of High-Level Nuclear Waste Disposal in the United States," in *Siting by*

Choice: Waste Facilities, NIMBY, and Volunteer Communities, Don Munton, ed., (Georgetown University Press: Washington D.C., 1996): 108-41.

(Macfarlane, 2001) Macfarlane, Alison. "Interim Storage of Spent Fuel in the United States," *Annual Review of Energy and the Environment*, vol. 26, 2001: 201-35.

(Meserve, 2002a) Meserve, Richard A. Letter to Congressman Edward J. Markey, 4 March 2002 [http://www.house.gov/markey/ISS_nuclear_ltr020325a.pdf]

(Meserve, 2002b) Meserve, Richard A. "Nuclear Power Plant Security," Statement submitted by the NRC to the Committee on Environment and Public Works, U.S. Senate, 5 June 2002 [http://www.nrc.gov/reading-rm/doc-collections/congress-docs/congress-testimony/2002/ml021570116.pdf]

(NAS, 1994) "Building Consensus Through Risk Assessment and Management of the Department of Energy's Environmental Remediation Program," Committee to Review Risk Management in the DOE's Environmental Remediation Program, National Academy of Sciences, National Research Council, National Academy Press, 1994 [http://www.nap.edu/openbook/NI000191/html/R1.html]

(NAS, 2005a) "Developing 'Risk-Based' Approaches for Disposition of Transuranic and High-Level Radioactive Waste," National Academy of Sciences, Current Projects, BRWM-U-03-01-A [http://www.nas.edu] · The published report, *Risk and Decisions About Disposition of Transuranic and High-Level Radioactive Waste*) is available at [http://newton.nap.edu/catalog/11223. html#toc].

(NAS, 2005b) "Safety and Security of Commercial Spent Nuclear Fuel Storage," Committee on the Safety and Security of Commercial Spent Nuclear Fuel Storage, Board on Radioactive Waste Management, Division on Earth and Life Studies, National Academy of Sciences, National Research Council, National Academy Press, 2005 [http://www.nap.edu/openbook/0309096472.html]

(NEI, 2004a) "Powering Tomorrow…With Clean, Safe Energy," Vision 2020, Nuclear Energy Institute [http://www.nei.org/documents/Vision2020_Backgrounder_Powering_Tomorrow.pdf]

(NEI, 2004b) "Post-Sept. 11 Security Enhancements: More Personnel, Patrols, Equipment, Barriers," Nuclear Energy Institute, Safety and Security report [http://www.nei.org/index.asp?catnum=2&catid=275]

(NEI, 2004c) "Safe Interim On-Site Storage of Used Nuclear Fuel," Nuclear Energy Institute, Nuclear Waste Disposal report [http://www.nei.org/index.asp?catnum=2&catid=70]

(NRC, 1996) "Strategic Assessment Issue: 6. High-Level Waste and Spent Fuel," Nuclear Regulatory Commission, 16 September 1996 [http://www.nrc.gov/NRC/STRATEGY/ISSUES/dsi06isp.htm]

(NRC, 2001) *Disposition of High-Level Waste and Spent Nuclear Fuel: The Continuing Societal and Technical Challenges*, Nuclear Regulatory Commission, Committee on Disposition of High-Level Radioactive Waste Through Geological Isolation, Board on Radioactive Waste Management, Division on Earth and Life Sciences (National Academies Press: Washington D.C., 2001).

(OCEP, 2000) "Owl Creek Energy Project: Private Interim Spent Fuel Storage Facility," Owl Creek Energy Project, Global Spent Fuel Management Summit, 3-6 December 2000.

(PFS, 2000) "The PFS Facility," Private Fuel Storage, LLC [http://www.privatefuelstorage.com/project/facility.html]

(PFS, 2003) "Private Fuel Storage Challenges Licensing Board Decision," Private Fuel Storage, News Release 3/31/03 [http://www.privatefuelstorage.com/whatsnew/newsreleases/nr3/31/03.html]

(POGO, 2002) "Nuclear Power Plant Security: Voices from Inside the Fences," 12 September 2002, Project on Government Oversight [http://www.POGO.org/p/environment/eo-020901-nukepower.html]

(Rogers and Rogers, 1998) Rogers, Kenneth A. and Donna L. Rogers. "The Politics of Long-Term Highly Radioactive Waste Disposal," *Midsouth Political Science Review*, vol. 2, 1998: 61-71.

(Rogers and Kingsley, 2004a) Rogers, Kenneth A. and Marvin G. Kingsley. "The Politics of Interim Radioactive Waste Storage: The United States," *Environmental Politics*, vol. 13, No. 3, September 2004: 590-611.

(Rogers and Kingsley, 2004b) Rogers, Kenneth A. and Marvin G. Kingsley. "The Transportation of Highly Radioactive Waste: Implications for Homeland Security," *Journal of Homeland Security and Emergency Management*, Volume 1, Issue 2, Article 13, 2004.

(Stanton, 2005) Lawrence Stanton, Deputy Director, Protective Security Division, Information Analysis and Infrastructure Protection, U.S. Department of Homeland Security, Interview, Washington D.C., 20 May 2005.

(Suplee, 1995) Suplee, Curt. "A Nuclear Problem Keeps Growing," *The Washington Post*, 31 December 1995: A1, A18, A19.

(White House, 2001) "National Energy Policy," White House, 2001 [http://www.whitehouse.gov/news/releases/2001/05/20010517-2.html]

Chapter 7

What to Do?

It is important to note that the problem of what to do with radioactive waste piling up at reactor sites is not unique to the United States. The issue of how best to properly dispose of highly radioactive waste is an international concern of growing importance. Commercial nuclear energy reactors across the globe (especially Europe) are generating substantial quantities of spent nuclear fuel. Today, there are over 440 nuclear power plants operating in 31 countries and global nuclear power generation accounts for approximately 16 percent of the world's electricity production. (IAEA, 2006: 1) Moreover, nuclear power production is expected to increase significantly over the next 25 years as more power plants come on-line in both industrialized countries and the developing world. Thus, the amount of highly radioactive waste generated will significantly increase in the coming years.

In spite of need to dispose of highly radioactive waste properly, global efforts to find a solution to the radioactive waste problem have been slow to materialize. Thus, similar to the United States, generally a de facto policy of storing the waste on-site has evolved. While this approach clearly will not be a viable policy for the long term, it has been difficult to forge a consensus on what to do next. For example, the first non-U.S. radioactive waste repository is not expected to be operational until sometime around 2020. As of now, no country has yet solved the problem of what to do the mounting inventories of spent nuclear fuel created as a by-product of nuclear electrical power generation or the high-level waste from nuclear weapons production.

In the United States, the issue of the disposal of highly radioactive waste generated minimal concern in the early years after the discovery of nuclear fission since it was not perceived to be an issue that required immediate attention. Rather, the disposal of nuclear waste generally was considered to be a problem that could be addressed sometime in the future. This mindset delayed recognition of the impending problem of how best to cope with the mounting inventories of nuclear waste. This, in turn, prevented any real progress on developing a long-term solution to the problem since it essentially allowed policy makers to avoid the issue until it became a more pressing and immediate problem. Even after it became evident that the disposal problem of highly radioactive waste demanded more immediate action, effective policies to develop a long-term solution to the problem have been rather slow to materialize.

The slow progress in finding a long-term solution has been due primarily to the fact that disposal has been the most contentious alternative in dealing with radioactive waste. The issue of highly radioactive waste disposal is inherently conflictual for a number of reasons. First, the costs of managing the nuclear waste are tangible (i.e., readily apparent) and substantial, while the benefits of finding a solution, such as enhancing security, are intangible (i.e., not readily apparent). Second, the costs incurred for disposal are apparent immediately, while the benefits will be realized sometime in the future. Third, while the benefits of achieving a long-term solution to the problem are dispersed to the country as a whole, the costs are concentrated in the locale that is saddled with hosting the disposal facility. For these types of issues, policymakers generally prefer to defer decisions to a later date in order to avoid conflict. However, it is imperative that some equitable resolution of the problem be attained in the near-term so that future generations do not bear the safety, health and economic costs of disposing of the highly radioactive waste that has been created by a previous generation.

The problem of highly radioactive waste storage and disposal has become one of the most controversial aspects of nuclear technology. Unfortunately, as the need for action has become more acute, conflict generated by the politicization of the issue has delayed progress on developing a long-term solution to the problem. As the years passed, however, spent nuclear fuel from commercial nuclear power generation and high-level waste from defense related weapons production continued to accumulate at reactor sites and defense facilities across the country. Today, homeland security and energy security concerns associated with the accumulating nuclear waste have prompted policy makers to pay additional attention to the issue.

The events of 9/11 have focused increased attention on homeland security issues due to the vulnerability of storing highly radioactive waste on-site at numerous locations around the country. And clearly, storing large quantities of highly radioactive waste near major population centers heightens anxiety over a possible terrorist attack with the potential for a significant release of radiation. Since millions of Americans live within a few miles of the stored nuclear waste, it would appear to be prudent to minimize the number of potential targets by removing as much of the on-site spent nuclear fuel and high-level waste as possible.

At the same time, energy security concerns began to surface after it was revealed that a number of commercial nuclear power facilities might have to be shut down prematurely due to a lack of sufficient on-site storage capacity. Concerns over adequate energy supplies have reinforced the perception of the need for the United States to reduce its dependency on foreign sources of energy. However, this can only be done if the United States does not lose any of its current domestic energy generation capacity and is able to keep pace with future energy demands. Since nuclear power now accounts for about one-fifth of total U.S. electricity generation, and electricity demands are projected to increase by about 50 percent over the next twenty years, it is clear that the U.S. can not afford to lose nuclear power generated electricity prematurely due to a lack of adequate on-site storage.

In spite of these homeland security and energy security concerns, however, progress has been slow in forging a long-term solution to the problem of what to do with spent nuclear fuel and high-level waste. The lack of resolution of this problem has been due primarily to the politicization of the issue. As The Office of Civilian Radioactive Waster Management points out, "The safe disposal of radioactive waste is not only a complex scientific and engineering problem, but also a difficult social and political issue." (OCRWM, 2004: 1)

Since the current on-site storage regime has had relatively little conflict, it has been easy for policy makers to support the status quo. Continuing a relatively non-controversial status quo policy, especially one that has been in place for a number of years, involves minimal short-term political risk. On the other hand, confronting a controversial issue head-on is more difficult since it involves making decisions that are potentially contentious and politically risky. Thus, a centralized interim storage or a long-term disposal option will require the political will to overcome the inertia of continuing on-site storage. Thus, politics plays an important role in formulating policy for controversial issues such as highly radioactive waste.

Factors Affecting U.S. Radioactive Waste Policy

The structure of the U.S. political system almost guarantees that coping with controversial issues such as radioactive waste storage or disposal will be difficult and fraught with conflict. The American political system was designed by the Founders to ensure that no single entity would be able to accumulate too much power at the expense of the people. The principles of limiting the power of the national government (e.g., federalism) and diffusing the power granted to the national government (e.g., separation of powers) are a foundation of the American political system. While this structure has prevented an abuse of power by the federal government, it also has ensured that policymaking – especially for contentious issues such as highly radioactive waste – would be slow and arduous.

Federalism has almost ensured a certain amount of conflict will occur between the federal government and the states. Clearly, the federal government believes that the national security and energy security considerations associated with the mounting quantities of highly radioactive waste trump state objections over hosting a centralized interim storage facility or permanent repository. However, the states – and especially the western states since the majority of nuclear waste is generated in the East – bristle at the suggestion that the federal government has the power to site a nuclear waste facility within state borders over a state's objections. Thus, the inherent tension between the states and the federal government over the issue of what to do with highly radioactive waste invites conflict.

The states, tribal governments, and local governments clearly can play an important role in the radioactive waste policy debate since they can be involved either as a host for a storage or disposal facility, or are involved with the transportation of highly radioactive waste through their jurisdictions. In essence, however, the states generally do not want to host a centralized interim storage facility or a permanent repository due to the political fallout that such a facility would generate. On the other hand, the federal government understands that the homeland security and energy concerns associated with having highly radioactive waste stored at numerous locations across the country dictates that some type of centralized solution is necessary. Thus, the stage is set for significant conflict between the federal government and the states identified to host a centralized highly radioactive waste storage or disposal facility. While the principle of national supremacy ensures that any attempts by states to regulate radioactive waste storage and disposal must not contradict federal regulations, it does not ensure there will not be conflict. In fact, the likelihood of conflict between the federal government and the states prompted federal officials (and later the nuclear utilities) to negotiate with Native American tribes to build and operate an interim radioactive waste storage facility. The use of Native American tribal lands is an interesting choice since the use of tribal lands could circumvent state sovereignty and erode a state's ability to influence the decision making process.

The separation of powers between the legislative, executive, and judicial branches creates a system of checks and balances. This also ensures that devising any radioactive waste policy will be slow and difficult since all three branches have to concur on a policy solution. Clearly, the president sets the tone by which policies are proposed. For example, by recommending Yucca Mountain to Congress for a national repository, President Bush moved the process forward. In addition, the president can veto proposed radioactive waste legislation as President Clinton did in the 1990s. Although Congress has the ability to override the veto with a two-thirds majority in both chambers, it is unlikely this would happen since only about 7 percent of presidential vetoes are overridden successfully. While the president can propose policy such as recommending Yucca Mountain for the national repository, it is Congress that has the ultimate responsibility for passing the legislation to establish and fund the facility. For example, Congress passed the 1982 Nuclear Waste Policy Act outlining the process for establishing a national repository and then in 1987 amended the legislation singling out Yucca Mountain, Nevada, as the only site to be studied. Even if the executive and legislative branches agree on a particular course of action, the judicial branch can block action on any radioactive waste initiative by virtue of exercising judicial review which allows the courts to declare executive and legislative acts as unconstitutional and thus, null and void. Since highly radioactive waste litigation has increased substantially in recent years, the federal courts are now more active in the radioactive waste policy debate. In fact, the courts have now become the primary mechanism used by opponents of either a centralized interim storage facility or a permanent repository to block or derail highly radioactive waste storage or disposal initiatives.

A number of federal bureaucratic agencies have the responsibility and jurisdiction over highly radioactive waste issues. While a variety of government agencies are involved in some aspect of nuclear regulation, there are three primary federal organizations tasked with managing spent nuclear fuel and high-level waste: the U.S. Department of Energy (DOE), the U.S. Nuclear Regulatory Commission (NRC), and the U.S. Environmental Protection Agency (EPA). The DOE is responsible for promoting nuclear energy, cleaning up high-level waste at DOE weapons facilities, managing the National Transportation Program, and developing a permanent disposal capacity to manage highly radioactive waste. The NRC is responsible for regulating reactors, nuclear materials and nuclear waste, and developing regulations to implement safety standards for licensing a national highly radioactive waste repository. The EPA is the primary federal agency charged with developing radiation protection standards; providing guidance and training to other federal and state agencies in preparing for emergencies at U.S. nuclear plants, transportation accidents involving shipments of radioactive materials, and acts of nuclear terrorism; and developing the environmental standards to evaluate the safety of a national geologic repository. Moreover, there are additional federal actors that also play an important role in highly radioactive waste policy. For example, the U.S.

Department of Transportation is involved in radioactive waste transportation issues, the Federal Emergency Management Agency would play a role in the response and recovery phase of a nuclear waste incident, and the Nuclear Waste Technical Review Board (NWTRB) is an independent U.S. agency of the U.S. government tasked to provide independent oversight of the U.S. civilian radioactive waste management program.

As a pluralist democracy, competing groups and shifting alliances help to determine policy. As a result, the variety of interest groups involved in the radioactive waste debate ensures that politics will play a major role in the formulation and implementation of highly radioactive waste policy. Numerous interest groups have been active in the radioactive waste policy debate. They generally can be broken down into three major categories: (1) state and local government affiliated organizations; (2) environmental and citizen activist groups; and (3) the commercial nuclear power industry and supporters. These groups are active in lobbying the federal and state governments on radioactive waste storage and disposal policy. State and local government affiliated organizations have focused on emergency response to a potential accident involving radiological materials. Environmental and citizen activist groups argue that the radioactive waste can continue to be stored on-site at the point of origin, thus alleviating the need to construct a storage or disposal facility as well as transport it to a centralized facility. In contrast, the commercial nuclear power industry has been a strong advocate of building either a centralized interim storage facility or a national repository for highly radioactive waste in order to avoid the premature shutdown of the reactors due to a lack of sufficient on-site radioactive waste storage capacity.

The media also can play an important role in the nuclear waste policy debate by virtue of its gatekeeper role of determining what to report as news, and thus shaping citizen perceptions. If the media decides to report on a controversial issue, then it will have a high likelihood of being put on the political agenda. For example, the extensive media coverage of the Three Mile Island and Chernobyl reactor accidents focused considerable attention on the safety of commercial nuclear power plants. As a result, public opposition to nuclear generated power rose significantly, not only preventing the construction of new plants, but also ultimately leading to the closing of several commercial nuclear power facilities. In more recent years, there has been increased national coverage of the Yucca Mountain (Nevada) repository. Moreover, there has been a substantial amount of state and local reporting on the Skull Valley (Utah), and to a lesser extent Owl Creek (Wyoming), centralized interim storage initiatives. The media could play an important future role in the debate either in a positive or negative way. Inaccurate or unfair negative reporting could serve to undermine public confidence, which would likely increase citizen opposition to any storage or disposal plan. For example, according to a Department of Homeland Security (DHS) official, a good deal of information in the public and media is based on erroneous assumptions such as the highly radioactive waste on-site could cause a nuclear explosion. And, when agencies such as the DHS try to explain differences between perceptions and reality, they are labeled as "being disingenuous." (Stanton, 2005) On the other hand, if the media focuses on the positive safety record and the low likelihood of an accident that could result in a release of radiological material, then citizen opposition will likely be minimized.

Over the years the issue of highly radioactive waste storage and disposal has not been an especially salient issue with the public. Thus, there has been little public pressure on either Congress or the White House to act. This, in turn allowed policy makers to put off even addressing the issue. Even after it became apparent that the issue of highly radioactive waste could no longer be deferred, policy makers were reluctant to develop a long-term policy to cope with the ever-increasing quantities of radioactive waste being generated.

The question is why the highly radioactive waste storage and disposal issue has been such a low priority with policy makers even after it became evident that something had to be done. There are a number of possible explanations. First, the safety record for highly radioactive waste storage on-site has been good. There have not been any major incidents that have garnered the headlines which would induce panic or fear. Second, the focus on nuclear facilities always has been on the reactors and their safety. The Three-Mile Island incident in 1973 and the Chernobyl accident in 1986 received considerable media coverage which served to focus even more attention on reactor safety, but not the nuclear waste stored on-site. Third, the highly radioactive waste storage and disposal issue is highly technical and beyond the understanding of most of the public. Complex technical issues do not normally generate much public interest. Fourth, coping with the problem of what to do with the waste is costly. It is expensive to build a storage or disposal facility. Moreover, the costs are immediate and readily apparent while the benefits are difficult to measure and long-term. In an era of limited resources, it is easier for policy makers to defer an issue that is expensive, especially when the benefits are long-term and not readily apparent. Fifth, since highly radioactive waste storage and disposal has not been a salient issue with the general public, there has been little pressure on either Congress or the White House to act. Policy makers tend to pay more attention to issues that are a high priority for the public. Issues that are not a priority tend to be deferred to a later date when the problem becomes more apparent. Finally, attempts to address the highly radioactive waste issue are politically charged, primarily due to opposition by the states identified as a potential host. The not-in-my backyard (NIMBY) phenomenon is a difficult hurdle to overcome. Because the U.S. political system tends to operate on consensus and compromise, divisive issues such as highly radioactive waste storage and disposal tend to be avoided, if possible, at all levels of government.

What to Do?

The slow pace in coping with highly radioactive waste storage and disposal has had some positive aspects. First, as the years have passed, technology has advanced which allows for a more technically sophisticated solution to the problem. Second, the additional time provides a better opportunity to build confidence among affected parties, which is especially important in a democracy where compromise and consensus are an important principle of the political system. One of the fundamental aspects of a liberal democracy is the concept of a "social contract" between those who govern (i.e., elected officials) and those who are governed (i.e., the citizens). Thus, the consent of the governed is an important foundation of democracy. This is especially true for nuclear related issues where public fears can be exaggerated.

On the other hand, the delay in constructing a national repository for the disposal of spent nuclear fuel and high-level waste has potentially significant negative repercussions for homeland security and energy security. Storing large quantities of highly radioactive waste at numerous locations around the country heightens anxiety over the possibility of a terrorist attack with a potentially significant release of radiation. While there are differences of opinion over the vulnerability of the on-site waste, it is clear that the spent fuel cooling pools and dry casks are more vulnerable than the reactors which are housed within concrete and steel reinforced buildings. The energy security implications are more evident. Unless a centralized storage option or a permanent disposal solution are found in the near future, some commercial nuclear power facilities will be forced to shut down their reactors prematurely due to a lack of sufficient on-site storage space for the spent nuclear fuel.

So, what are the alternatives? Basically, there are three options (other than reprocessing) in dealing with highly radioactive waste: continue the current on-site storage regime, find an acceptable centralized interim storage solution, or, achieve an acceptable permanent disposal solution. The major problem is that while homeland security and energy security concerns would mandate that either a centralized interim storage or permanent disposal solution should be pursued, the political realities have made either of these options difficult. Thus, the on-site storage regime has remained the primary focus.

Should policy makers continue storing spent nuclear fuel from commercial nuclear facilities and high-level waste from defense plants on-site in order to avoid the negative political ramifications associated with any centralized storage or disposal initiatives? The continued on-site storage of highly radioactive waste at numerous locations around the country (some near large metropolitan centers) presents a potentially unacceptable homeland security risk due to the potential for a terrorist attack, sabotage, or theft of the material. In 2005, the National Research Council released a report that questioned the safety of storing spent nuclear fuel on-site. The report stated that in spite of the fact that it would be difficult to attack a nuclear power facility; the radioactive waste on-site would be an attractive target for terrorists due to the potential for a significant release of radiation. While the report specifically singled out spent fuel pools as being vulnerable to a potential terrorist, it also stated that dry cask storage also was vulnerable to some types of terrorist attacks. (NAS, 2005: 6-7) While it would be difficult to provide additional protection for the spent fuel pools, the more vulnerable dry casks could be moved from above ground to a more protected area on-site until either a centralized interim storage facility or a permanent national repository is ready to accept highly radioactive waste. Even if the most vulnerable waste is moved to a shielded area, the current on-site storage regime still does not adequately address homeland security concerns and thus, is assessed as poor. In addition, the current on-site storage regime provides poor energy security due to the potential for having to shut down some commercial nuclear energy reactors at facilities where there is a lack of adequate on-site spent fuel storage space. The possible premature closing of commercial nuclear power facilities due to a shortage of adequate on-site spent nuclear fuel storage poses an unacceptable energy security risk (i.e., poor) during a time when the U.S. cannot afford to increase its dependency on foreign sources of energy.

Conflict associated with the current on-site storage regime has been relatively mild and involves only minimal short-term political risk. Since policymakers generally prefer to follow existing policy, or make only incremental policy adjustments, storing the waste on-site has been the preferred alternative. However, that does not imply that there has not been conflict over the on-site storage of highly radioactive waste – although the conflict has been local and sporadic. While conflict generally has been relatively mild, it would be reasonable to expect discord to increase in the future as on-site storage capacity reaches it limit. Thus, while the overall political risk associated with continuing temporary on-site storage currently is assessed to be low, the lack of resolution of this issue ultimately will increase conflict and political risk.

Should policy makers support establishing a centralized interim storage facility in case the Yucca Mountain project is further delayed or canceled? Clearly, centralized interim radioactive waste storage would address some homeland security concerns. While some spent nuclear fuel would remain on-site, the most vulnerable on-site waste could be moved to a more secure centralized facility. However, significant quantities of spent nuclear fuel would have to be transported to any centralized interim storage facility. Moreover, a centralized facility with a large quantity of aboveground storage would be an attractive terrorist target. Thus, security concerns associated with continued on-site dry cask storage as well as transportation concerns would lower the overall homeland security risk to fair. Energy security at a centralized interim storage facility would rate better since an interim storage facility would provide a temporary pathway for spent nuclear fuel, which would postpone the shutdown of reactors prematurely due to a lack of sufficient on-site storage space. However, since a centralized interim storage facility would only result in a partial fix for the mid-term (not the long-term) the overall energy security assessment for a centralized interim storage option would only be considered good.

Although a centralized storage option was first proposed over 30 years ago, conflict and politicization of the issue have prevented the construction of any centralized facility. Initial efforts by local jurisdictions to consider hosting a centralized interim storage facility generally were blocked by the host state. Thus, while a centralized storage option for highly radioactive waste would partially address both homeland security and energy security concerns, the politicization of the issue has produced considerable conflict in the states that would possibly host such a facility. As a result, the overall political risk associated with constructing a centralized interim storage facility for spent nuclear fuel is assessed to be moderate. The concept of centralized interim highly radioactive waste storage primarily has been controversial since opponents have voiced concern that any interim facility could become a de facto repository. Thus, any interim storage solution has been inextricably tied to progress on long-term disposal. While a centralized interim storage solution will address some of the more immediate problems associated with the continued on-site storage regime, it does not permanently solve the problem of what to do with the highly radioactive waste and offers only a temporary solution.

Should policy makers aggressively pursue the completion of a permanent repository as soon as possible due to the potential homeland security and energy concerns associated with the current on-site storage regime? According to the National Research Council, "After four decades of study, geological disposal remains the only scientifically and technically credible long-term solution available to meet the need for safety without reliance on active management." (NAS, 20001: 27) A national repository for the permanent disposal of highly radioactive waste (e.g., Yucca Mountain) would provide excellent homeland security since the nuclear waste would be stored below ground in a highly secured facility where the risk of sabotage or a terrorist attack would be minimal. However, a national repository would require transporting large quantities of highly radioactive waste across the country. Thus, while a secure national repository would provide excellent protection against a terrorist attack, the overall homeland security risk is assessed only as very good due to transportation security concerns. A national repository would provide a long-term pathway for spent nuclear fuel, alleviating the need to shut down commercial nuclear power reactors prematurely due to a lack of sufficient on-site nuclear waste storage space. Thus, overall energy security would be assessed as excellent.

However, the permanent disposal of highly radioactive waste has been the most politically contentious option primarily because the NIMBY reaction has generated opposition in states designated as a potential host for a nuclear waste repository. According to the NRC, geologic disposal has three main sources of uncertainties. First, the geologic properties of the site must be well identified and the future behavior of the storage facility must be well understood. Second, can the natural geologic processes jeopardize the efficacy of the repository? Third, can unforeseen human activities in the future jeopardize the safety of the site? (NAS, 2001: 26) Thus, these issues must be addressed prior to granting a license for the repository. Moreover, the site selection process was perceived as being unfair since Yucca Mountain, Nevada, was singled out to host the national repository because Nevada did not have the political influence of the other potential host states. Thus, resentment over the perceived unfairness of the site selection process coupled with apprehension over the geologic and environmental suitability of the Yucca Mountain site has generated additional controversy. As a result, the attempt to construct a geologic repository has become intensely politicized; and thus, the political risk is assessed as high.

While homeland security and energy security concerns should determine highly radioactive waste policy, in reality, it is the political considerations associated with each option that have been the driving force behind policy decisions. Thus, while security considerations would seem to dictate that a centralized storage or disposal option should be made available in the near-term, political realities have prevented this from happening. More importantly, these same political realities will clearly continue to be an obstacle to developing a long-term solution to mitigating both homeland security and energy security concerns associated with storing large quantities of highly radioactive waste at numerous locations around the country. Thus, the lack of political will to cope with the problem adequately means that over half of the U.S. population will continue to live for quite some time near locations where this lethal waste is stored.

Recommendations

There are a number of recommendations that would help to lessen the impact of the delay in achieving a coherent long-term policy on what to do with highly radioactive waste. While these recommendations will not solve the dilemma of how best to cope with the mounting inventories of highly radioactive waste, they can provide some relief in the near term by mitigating the impact of nuclear waste on both homeland security and energy security concerns.

Storage and Disposal

The realization that on-site storage does not provide a long-term answer to the problem of highly radioactive waste storage and disposal has intensified the search for a centralized interim solution. However, the fate of the interim storage of highly radioactive waste storage is still uncertain. On the one hand, the realization of the need to continue to produce nuclear energy have prompted efforts to ensure that no reactors will be forced to shut down prematurely due to a lack of sufficient on-site storage. On the other hand, it is clear that the process has become so contentious and politicized, that any centralized siting proposal will meet stiff resistance from the state where the proposed facility will be located.

A dual track strategy should be pursued. First, continue the quest for a long-term disposal solution. That entails continuing to pursue the Yucca Mountain initiative with the understanding that all environmental and geological concerns must be addressed forthrightly. A transparent process will lessen conflict and politicization of the issue. Second, it is paramount that a mid-term centralized storage option be pursued aggressively. In fact, in 2004, the National Commission on Energy Policy urged the Bush Administration and Congress to establish two centralized interim storage facilities at separate locations in order to provide interim back-up to a geologic repository program. (NCEP, 2004: 61) In the event that Yucca Mountain does receive final approval, it will be several years before the facility will be ready to accept nuclear waste. In the meantime, a centralized storage option can provide a temporary solution that will address some of the more immediate homeland and energy security concerns. At the same time, if the Yucca Mountain initiative does not go forward as planned, then there will be even more of a need to move the most vulnerable spent nuclear fuel and high-level radioactive waste to a more secure centralized facility. Thus, it would be prudent to pursue this dual track strategy in order to maximize the ability to at least provide a partial and temporary solution to the vexing problem of what to do with the highly radioactive waste being stored around the country.

Transportation

The key to enhancing security and muting much of the opposition to transporting highly radioactive waste is to convince state and local officials, as well as the attentive public, that not only is the risk of an incident low, but that there also has been sufficient attention and resources given to ensuring that the emergency response capability will be effective and timely. This can best be accomplished by following a two-track approach. First, provide the necessary training and equipment to state and local emergency first responders so they have the capability to respond quickly and effectively to any nuclear waste transportation incident.[1] Since the federal government has acknowledged that it will take possibly as long as 48 hours to respond to a radiological incident, federal officials must give state and local authorities sufficient time and resources to prepare adequately for the impending radioactive waste shipments. This can best be accomplished by adhering to the Nuclear Waste Policy Act which mandates the federal government provide adequate resources and training to state and local governments prior to the initiation of any spent nuclear fuel transportation program. The WIPP transportation program has been successful in providing assistance and training to state and local governments affected by transuranic waste shipments through their jurisdictions and could be used as a model for any centralized highly radioactive waste disposal or storage facility. Second, it will be imperative to begin a public education campaign addressing not only the steps that have been taken to prevent an accident, but also the efforts that have been taken to mitigate the effect of any potential incident. Because the transportation of radioactive waste generally has not been a salient issue with the U.S. public, there generally has been little real understanding of the complexities of this issue.

This lack of public understanding has made it easier for opponents to exploit any concerns over the safety of transporting spent nuclear fuel which clearly will complicate government efforts to convince the public that transporting spent nuclear fuel is safe and reliable. As Zacha (2003) points out, "You can be sure that in the next months and years, Nevada, helped by antinuclear activists, will be promulgating all sorts of disinformation about the dangers, the hazards, of nuclear waste transport." (Zacha, 2003: 4) For example, transportation opponents have used the terms "mobile Chernobyl" to accentuate the perceived dangers of transporting highly radioactive waste across the country.

Since the public generally has not been engaged in this topic, there has not been significant pressure exerted on elected officials to adopt a coherent public policy on the issue. However, this will change if a major incident involving a harmful release of radioactive material occurs – especially if the incident takes place in or near a populated area. Thus, prior to beginning any large-scale transportation operation, the federal government must implement a program that will provide adequate resources and training to state and local emergency management officials, as well as educate the public on the risks of transporting spent nuclear fuel across the country.

---

1   Providing first responders the training and equipment to respond effectively to a radioactive waste transportation accident will also have the ancillary benefit of enhancing overall homeland security in the event of a general attack by terrorists.

In addition, the federal government could help to minimize risk and maximize public confidence by adhering to recommendations in a 2003 Government Accountability Office report on enhancing the security of transporting spent nuclear fuel. The GAO report concluded that the risk associated with transporting spent nuclear fuel could be reduced if the total number of shipments were minimized by using the mostly rail scenario. (GAO, 2003: 17) The report also stated that if the oldest spent fuel (i.e., least radioactive) were shipped first, the overall risk would be reduced. This generally would be the radioactive waste from on-site dry cask storage. In addition, GAO concluded that security of the radioactive waste shipments could be enhanced by using dedicated trains carrying only spent fuel since rail shipments would shorten the duration of the shipment by several days as well as ensure hazardous flammable materials are not shipped with the radioactive waste. (GAO, 2003: 22) These recommendations seem a prudent course of action which could reduce the vulnerability of transporting highly radioactive waste across the country, sometimes through heavily populated areas.

Concluding Thoughts

Measuring the ability of options to achieve desired goals is a complicated process due to the inherent assumptions that inevitably must be made. Nevertheless, a theoretical construct that outlines the basic fit of options and goals can be helpful in determining desired outcomes. The degree of goal achievement can be ranked from "near zero" to "near perfect." (Kugler, 2006: 76) While it may not be difficult to assess the current situation, estimating the eventual policy outcome can be a more difficult undertaking since projecting the future always has some uncertainty. This is even truer when politics plays a central role in decision making since the rational choice may not be the chosen option (generally conflict is avoided at all costs). In spite of these inherent deficiencies, it can be helpful to assess different policy options and the likely outcome (see Figure 7.1 – Measuring Ability of Options to Achieve Desired Homeland Security Goals and Figure 7.2 – Measuring Ability of Options to Achieve Desired Energy Security Goals).

	Poor	*Fair*	*Good*	*Excellent*
On-Site Storage	X			
Centralized Storage		X		
Permanent Disposal			X	

Figure 7.1 Measuring Ability of Options to Achieve Desired Homeland Security Goals: Degree of Goal Achievement

	Poor	*Fair*	*Good*	*Excellent*
On-Site Storage	X			
Centralized Storage			X	
Permanent Disposal				X

Figure 7.2 Measuring Ability of Options to Achieve Desired Energy Security Goals: Degree of Goal Achievement

Clearly the current on-site storage regime does not adequately address either homeland security or energy security concerns. The centralized interim storage option partially addresses both homeland security and energy security – at least for the near term. The permanent disposal option would address both homeland security and energy security for the long-term. However, the level of political conflict is a driving force in policy formulation and avoiding political conflict is generally the desired goal (see Figure 7.3 – Measuring Political Conflict).

	Prefer/Low	*Moderate*	*High/Avoid*
On-Site Storage	X		
Centralized Storage		X	
Permanent Disposal			X

Figure 7.3 Measuring Political Conflict: Degree of Conflict

Because policy makers will gravitate towards the path of least resistance, it is likely that the on-site alternative will be the preferred option until it is no longer viewed to be a viable strategy. Thus, in spite of the negative security ramifications associated with on-site storage it will be difficult for elected officials to pursue either a centralized interim storage option or permanent disposal in a deep geologic repository. The basic problem with the highly radioactive waste issue is that the best long-term solution to the problem (i.e., disposal) is the most contentious politically, while the least desirable long-term option (i.e., continued on-site storage) is the most palatable politically since it involves the least political risk (see Figure 7.4 – Political Conflict v. Security). So, the basic problem remains, where do we go from here?

	Low	*Moderate*	*High*
On-Site Storage	Conflict & Security		
Centralized Storage		Conflict & Security	
Permanent Disposal			Conflict & Security

Figure 7.4 Political Conflict v. Security

The failure to act decisively has generated a problem of almost crisis proportions. As James Lee Witt, former director of the Federal Emergency Management Agency has stated: "Sometimes a crisis *becomes* a crisis simply because someone has failed to act." (Witt, 2002: 1, emphasis original) This clearly has been the case with the delay in developing an effective long-term policy in dealing with highly radioactive waste. Any further delay completing either a centralized interim storage facility or a national repository places an unacceptable burden on national security and energy security. In today's uncertain world, the potential vulnerabilities associated with the current on-site radioactive waste storage regime are too great to maintain the status quo. While there is controversy over the exact extent of the vulnerability of the on-site storage method, the potential consequences of a major incident would dictate moving the most vulnerable highly radioactive waste to a more secure environment.

Homeland security and energy security concerns that are associated with the ever-increasing amounts of highly radioactive waste should trump political considerations. At the same time, however, it must be acknowledged that political considerations always will be an important determinant of policy. Thus, it will be important for elected officials not to undercut the repository project by providing inadequate funding or allowing Nevada to block, or unnecessarily delay, constructing and operating a permanent repository for highly radioactive waste (provided the Yucca Mountain site is found to be suitable). This means that Congress will have to provide adequate funding in the future for the Yucca Mountain project, something it has not done in the past. In prior years, the Nevada congressional delegation has had considerable success in reducing the budgets for the Yucca Mountain repository. This, in turn, has delayed the project. While the opposition of Nevada is understandable, other members of Congress should not allow a minority to block a project that would be beneficial for overall U.S. homeland security and energy security. At the same time, Congress has raided the Nuclear Waste Fund which was established to provide the necessary resources to fund a national repository for highly radioactive waste. This has placed an unacceptable burden on funding for the repository since it takes away monies that already were earmarked for the disposal of spent nuclear fuel. In fact, in 2004, the National Commission on Energy Policy urged Congress to restore monies taken from the Nuclear Waste Fund. (NCEP, 2004: 60)

If the Yucca Mountain site is determined to be unsuitable for either geological or environmental reasons, then the project should be canceled. But, at this late date, these should be the only reasons that the project should be further delayed or canceled. Considering that Nevada hosted the country's nuclear weapons testing program, it might seem unfair that the state again would be selected, over its objection, to host the first national repository for highly radioactive waste. Nevada could argue that it has had to endure more than its share of sacrifices for the country. However, if the Yucca Mountain site is found to be suitable from a geological and environmental perspective, then it would seem rational to locate the repository in a sparsely populated area that has a history with nuclear technology (Nevada Test Site). It would appear to be a form of substituting one hazard for another. Economic incentives could be added that would help to soften the blow of having to host the repository.

Clearly, opponents of either a centralized interim storage facility or a permanent repository have legitimate concerns. Health and safety worries associated with a possible radioactive release, problems with the equity of concentrating so much waste in a community that did not reap the benefit of the activity that generated the waste, and concerns over a potential economic loss due to the stigma associated with hosting such a hazardous facility are issues that merit close scrutiny. Nevertheless, opponents of any centralized option clearly prefer the status quo of on-site storage at the point of generation; and it will be difficult to satisfy them since they will be content only if the highly radioactive waste storage or disposal facility is not built within their jurisdiction.

In the final analysis, the needs of the many should outweigh the needs of the few. That is, the national security and energy security concerns that potentially could impact the country as a whole should outweigh the concerns of the state where a centralized interim radioactive waste storage facility or permanent repository would be located. While this might seem unfair to the state or community where the nuclear waste facility will be located, it is a necessary consequence of protecting the country as a whole. If terrorists were to attack the highly radioactive waste stored on-site, and this resulted in a significant release of radioactivity, there would be a public outcry over the inability of government to protect public health and safety. Policy makers most certainly would then be forced to act. Why not take the necessary steps now to reduce vulnerabilities of the nuclear waste prior to an incident that would put American citizens at risk?

References

(GAO, 2003) "Spent Nuclear Fuel: Options Exist to Further Enhance Security," General Accounting Office, Report to the Chairman, Subcommittee on Energy and Air Quality, Committee on Energy and Commerce, U.S. House of Representatives, GAO-03-426, July 2003 [http://www.gao.gov/new.items/d03426.pdf]

(IAEA, 2006) "Nuclear Power and Sustainable Development", International Atomic and Energy Agency, April 2006 [http://www.iaea.org/Publications/Booklets/Development/npsd0506.pdf]

(IEER, 1999) "International Repository Programs," Institute for Energy and Environmental Research, May 1999 [http://ieer.org/sdafiles/vol_7/7-3/repos.html]

(Kugler, 2006) *Policy Analysis in National Security Affairs: New Methods for a New Era*, Center for Technology and Security Policy, National Defense University (NDU Press: Washington D.C., 2006).

(NAS, 2001) "Disposition of High-Level Waste and Spent Nuclear Fuel: The Continuing Societal and Technical Challenges," Committee on Disposition of High-Level Radioactive Waste Through Geologic Isolation, Board on Radioactive Waste Management, Division on Earth and Life Sciences, National Academy of Sciences, National Research Council, National Academy Press, Washington D.C., 2001 [http://www.nap.edu/openbook/0309073170/html/R1.html]

(NAS, 2005) "Safety and Security of Commercial Spent Nuclear Fuel Storage," Committee on the Safety and Security of Commercial Spent Nuclear Fuel Storage, Board on Radioactive Waste Management, Division on Earth and Life Studies, National Academy of Sciences, National Research Council, National Academy Press, 1994 [http://www.nap.edu/openbook/0309096472.html]

(NCEP, 2004) "Ending the Energy Stalemate: A Bipartisan Strategy to Meet America's Energy Challenges", The National Commission on Energy Policy, December 2004 [http://64.70.252.93/newfiles/Final_Report/index.pdf]

(OCRWM, 2004) "Assessing the Future Safety of a Repository at Yucca Mountain," Office of Civilian Radioactive Waste Management, U.S. Department of Energy, CD-ROM, 1 March 2004.

(Stanton, 2005) Lawrence Stanton, Deputy Director, Protective Security Division, Information Analysis and Infrastructure Protection, U.S. Department of Homeland Security, Interview, Washington D.C., 20 May 2005.

(Witt, 2002) Witt, James Lee. *Stronger in the Broken Places* (Times Books, NY: 2002).

(Zacha, 2003) Zacha, Nancy J. "Fear of Shipping," Editor's Note, *Radwaste Solutions*, March/April 2003 [http://www2.ans.org/pubs/magazines/rs/pdfs/2003-3-4-2.pdf]

Glossary

1972 London Convention on the Prevention of Marine Pollution by Dumping of Wastes and Other Matter: International treaty that prohibits the dumping of radioactive waste into the ocean.

2001 Baltimore Tunnel accident: Accident which resulted in a major fire that lasted for four days. Cited by transportation opponents as a classic example of an accident that could result in a significant release of radiological material.

accident: Rail or vehicle accident involving the transport of highly radioactive waste that occurred during transit.

deep bore hole disposal: Would emplace the radioactive waste canisters deep underground. Similar to geologic disposal but has the advantage of emplacing the waste in extremely deep levels well below any depth that a mined repository could achieve.

Department of Energy (DOE): Promotes a diverse supply of energy (including nuclear energy), and is responsible for developing a permanent disposal capacity to manage the nation's highly radioactive waste.

Department of Transportation (DOT): Has primary responsibility for establishing and enforcing standards for the shipment (both highway and by rail) of highly radioactive waste.

design based threat guidelines: Guidelines requiring commercial nuclear power plants to implement new security plans that are capable of defending against potential threats to security.

disposal: The permanent placement of radioactive waste.

dry casks: Steel and concrete containers that use a passive ventilation method to enable the spent fuel rods to continue to cool slowly.

energy security: Protection of energy sources by preventing premature shut down of nuclear reactors.

Environmental Protection Agency (EPA): Is the primary federal agency charged with developing radiation protection standards for protecting the public and the environment from harmful and avoidable exposure to radiation.

environmental justice: Siting undesirable projects in lower income communities that have reduced political influence (sometimes referred to as environmental racism).

excellent security risk: Addresses the security risk for the long-term.

excepted packaging: Packaging used to transport materials with extremely low levels of radioactivity. Does not require stringent guidelines for the radioactive waste container.

extraterrestrial disposal: Highly radioactive waste would be launched by rocket directly into the sun.

fair security risk: Partially addresses the security risk under consideration (e.g., may only be a partial or temporary fix).

Federal Emergency Management Agency (FEMA): Responsible for evaluating off-site emergency plans, and would play a crucial role in the response and recovery phase of any transportation accident involving nuclear waste that would result in the release of radioactivity.

general license: Granted to nuclear plant licensees to store spent fuel in NRC approved dry casks. A general license does not require the lengthy process of initiating a site specific license application requiring a safety analysis report, emergency plan, decommissioning plan, security plan, and environmental report.

geologic disposal: Consists of isolating the highly radioactive waste by emplacing the waste canisters in a mined underground repository.

good security risk: The security risk is generally addressed but with some reservations.

high-level waste: Highly radioactive waste generated by DOE as a result of nuclear weapons production (e.g., chemical reprocessing of spent nuclear fuel to recover plutonium), surplus plutonium from dismantled nuclear weapons, and defense research and development programs.

high political risk: Involves significant political conflict (e.g. substantial conflict existing at the national level).

highly radioactive waste: Describes both spent nuclear fuel generated from commercial nuclear power plants and high-level waste produced at DOE facilities.

Highway Route Controlled Quantity (HRCQ) shipment: Shipments must be clearly marked with the appropriate radioactive material warning sign, and preferred routing must be used which takes into account factors such as the potential accident rate along the transit route, the transit time, population density affected by the route, and time of day and day of the week.

highway transportation cask: Specially designed radioactive waste transportation cask for highway transport.

homeland security: Protection of the homeland from sabotage or terrorist attack.

ice sheet disposal: Waste would be emplaced in shallow boreholes where heat from the decomposing nuclear material would cause the waste canisters to sink slowly into the ice. The advantage of this method of disposal is that it would be located in a remote location.

incident: Any reported problem that occurred during any point of the shipment phase.

industrial packaging: Three levels of packaging (IP-1, IP-2, IP-3) used to transport materials which present a limited hazard. *IP-1* packaging must meet the same general design requirements for excepted packaging, *IP-2* packaging must pass free-drop and stacking tests, and *IP-3* packaging must also pass water spray and penetration tests.

Interim Storage Program: A federally owned and operated system for the interim storage of spent nuclear fuel) at one or more federally owned facilities.

interstate commerce clause: In the Constitution (Article I, Section 8), which gives the federal government control over activities that transcend state boundaries, to extend federal control over a number of areas that previously were under control of the states.

level 1 strain: Is where the strain on the cask's inner wall does not exceed .2 percent (the limit that stainless steel can withstand without violating its integrity).

level 2 strain: Is where the strain on the cask's inner wall does not exceed .2 percent-2 percent (stainless steel does not regain its shape and lead shielding may separate which may allow radiation to leak).

level 3 strain: Is where the strain on the cask's inner wall is between 2 percent-30 percent (can cause large distortions of the metal shape and the cracking of welded areas which may result in a serious release of radioactive material).

low political risk: Involves minimum conflict (e.g., only limited, generally localized conflict).

moderate political risk: Involves mid-level conflict (e.g., conflict is more evident, usually at the state level)

Monitored Retrievable Storage (MRS) concept: Where radioactive waste would be stored in special canisters above ground until ready for shipment to a permanent repository.

MRS Commission: Created to oversee implementation of the MRS concept and report on the viability of an interim radioactive waste storage facility.

national supremacy: Enshrined in the Constitution (Article VI) where federal laws "shall be the supreme law of the land." Thus, if a federal law and state law conflict, the federal law takes precedence over the state law.

NIMBY: Means "not in my backyard." A term used to describe opposition to siting a radioactive waste storage or disposal facility in the area.

notice of disapproval: Referred to as a "veto" in the press, a notice of disapproval is a formal notification by the state that it opposes the proposed facility. A notice of disapproval must be overridden by Congress.

Nuclear Regulatory Commission (NRC): Has overall responsibility for developing regulations to implement safety standards for licensing a national highly radioactive waste repository, and is responsible for regulating reactors, nuclear materials and nuclear waste through licensing requirements.

Nuclear Waste Policy Act of 1982 (NWPA): Began the process of establishing a deep, underground permanent national repository for highly radioactive waste. The 1982 Act required the Department of Energy (DOE) to first study multiple sites in the West, and identify a second site in the East for a repository.

Nuclear Waste Policy Amendments Act of 1987 (NWPAA): Amended the 1982 NWPA and singled-out Yucca Mountain, Nevada, as the only site to be studied.

Nuclear Waste Technical Review Board (NWTRB): Tasked to provide independent scientific and technical oversight of the U.S. program for the management and disposal of spent nuclear fuel and high-level radioactive waste and review the technical and scientific validity of the Department of Energy's activities associated with investigating the site and packaging and transporting wastes.

ocean/sub seabed disposal: One proposal advocated dropping canisters filled with nuclear waste into the deepest parts of the ocean such as the Marianas Trench. A second option proposed dropping missile-shaped steel waste canisters from the surface that would penetrate up to 70 meters (approximately 229 feet) deep into the seabed sediment. A third proposal advocated dropping the waste into deep pre-drilled boreholes in the ocean floor.

Office of Civilian Radioactive Waste Management (OCRWM): Tasked to develop and manage the federal highly radioactive waste system. The OCRWM has the primary responsibility for determining the suitability of the Yucca Mountain project.

Office of the Nuclear Waste Negotiator: Office charged primarily with promoting potential sites for interim storage.

on-site storage: Spent nuclear fuel stored on-site in spent fuel pools and dry cask containers.

Owl Creek Energy Project: A wholly private venture that proposed constructing a centralized interim storage facility near Shoshoni, Wyoming. The facility design comprised about 100-200 acres, and would be located within a privately owned section of undeveloped land approximately 2700 acres in size.

Plan B: A proposal by the State of Utah that would construct and operate a nuclear waste facility on state school trust lands.

poor security risk: Means that the storage/disposal option does not adequately address potential homeland security or energy security risks.

Price-Anderson Act: Provides liability protection for reactor operators in the event of an accident that results in a catastrophic release of radiation.

Private Fuel Storage, LLC: A consortium of eight utilities (Xcel Energy, Genoa Fuel Tech, American Electric Power, Southern California Edison, Southern Nuclear Company, First Energy, Entergy, and Florida Light and Power) that concluded a lease with the Skull Valley Goshute Indians to build a centralized interim highly radioactive waste storage facility.

proliferation: The spreading of nuclear weapons to other countries.

Radiological Assistance Program (RAP): Established to assist state and local governments to respond to an accident with a serious release of radioactive materials.

Radiological Assistance Team: Specially trained federal response team that can assist in the clean up of a radioactive waste accident.

Radioactive Materials Incident Report (RMIR) database: Database maintained by Sandia National Laboratory that compiles reports involving loss, exposure to, or release of radioactive materials incidents associated with radioactive materials.

rail transportation cask: Specially designed radioactive waste transportation cask for rail transport.

Regional Coordinating Offices (RCO): Eight regional offices established to coordinate a clean up response to a radioactive waste spill. The RCO could dispatch a Radiological Assistance Team to assist in the clean up.

remote uninhabited island disposal: Method of disposal where the waste would be far away from inhabited areas.

reprocessing (or partitioning): Reprocessing spent nuclear fuel and high-level waste in order to recover plutonium. Expected to reduce the inventories of spent nuclear fuel and the amount of time that the nuclear waste would remain a threat to health and the environment.

re-racking: Has increased spent nuclear fuel storage capacity by packing spent fuel rods closer together in cooling pools in order to increase spent nuclear fuel storage capacity.

Retrievable Surface Storage Facility (RSSF) concept: A centralized storage facility on the surface where the radioactive waste canisters could be retrieved easily for transfer to a permanent repository.

risk: Defined as the potential for a possible event to occur, coupled with the possible consequence of that event.

rod consolidation: Where the fuel rods are removed from the spent fuel assembly and rearranged more compactly inside a metal canister in order to increase spent nuclear fuel storage capacity.

rolling stewardship: Where the risk of managing the radioactive waste will "roll over" from generation to generation.

site-specific license: Allows the nuclear utility to store waste off-site (away from the reactor) as well as accept spent fuel from other plants.

Skull Valley (centralized interim storage facility): Proposed interim storage facility that is located about 45 miles southwest of Salt Lake City and would be approximately 100 acres in size. The complex would be able to store (above ground) up to 4,000 dry casks containing about 40,000 metric tons of spent fuel.

spent fuel pools: Concrete and steel lined pools located near the reactor and designed to provide radioactive shielding and cooling for the spent fuel rods.

spent nuclear fuel: The by-product of the fission process from reactors in commercial nuclear power facilities, nuclear submarines and ships, and university and government research facilities. The spent nuclear fuel is a highly radioactive waste that contains plutonium as well as unconsumed uranium.

storage: The temporary placement of radioactive waste.

strain: Is the stress or damage on a radioactive waste cask under different accident scenarios. There are three levels of "strain."

transmutation: Involves reprocessing the spent nuclear fuel by altering the radioactivity of the elements so that they remain a radioactive threat for a shorter period of time.

Transportation Emergency Preparedness Program (TEPP): Established to assist state and local governments to respond to an accident with a serious release of radioactive materials. TEPP integrates transportation and emergency preparedness activities and could serve as a model for the transportation of commercial spent nuclear fuel.

transuranic waste: Radioactive waste (i.e., heavier than uranium) that consists mainly of contaminated protective clothing, rags, tools and equipment, chemical residues and scrap materials resulting from defense activities such as nuclear weapons production. Transuranic waste is disposed of at the Waste Isolation Pilot Plant (WIPP).

Type A packaging: Container used to transport radioactive materials with higher concentrations of radioactivity. Type A packaging must be able to withstand a series of tests (penetration, vibration, compression, free drop, and water spray) designed to ensure that the container can survive an accident without the release of radioactive material.

Type B packaging: Used to transport materials with the highest radioactive levels. Type B packaging must pass a rigorous sequence of tests (free drop – from 30 feet, puncture – 40 inch drop onto a steel rod, heat – exposure to 1,475 degrees for 30 minutes, immersion – immersed under 3 feet of water, and crush-dropping an 1,100 pound mass from 30 feet onto the container) designed to ensure that the container can survive an accident without the release of radioactive material.

Bibliography

ABC News OnLine, "Report urges waste dump plan be abandoned," 18 February 2004 [http://www.abc.net.au/news/newsitems/s1047286.htm]

Abraham, Spencer. "A Review of the President's Recommendation to Develop a Nuclear Waste Repository at Yucca Mountain, Nevada," Secretary of Energy, Prepared Witness Testimony, Committee on Energy and Commerce, Subcommittee on Energy and Air Quality, 18 April 2002 [http://energycommerce.house.gov/107/hearings/04182002Hearing505/Abraham859print.htm]

Abraham, Spencer. Letter to President Bush recommending approval of Yucca Mountain for the development of a nuclear waste repository, Secretary of Energy, U.S. Department of Energy, 14 February 2002.

Abrahms, Doug. "As Yucca project stalls, Utah nuke waste dump hits fast track," *Reno Gazette Journal*, 4 April 2005 [http://www.rgj.com/news/stories.html/2005/04/02/96157.php]

Abrahms, Doug. "Reid cuts Yucca Mountain funding," *Reno-Gazette Journal*, 1/21/2003 [http://www.rgj.com/news/stories/html/2003/01/21/32573.php]

BMU, *Environmental Policy: Joint Convention on the Safety of Spent Fuel Management and on the Safety of Radioactive Waste Management*, Report Under the Joint Convention by the Government of the Federal Republic of Germany for the Second Review Meeting in May 2006, Federal Ministry for the Environment, Nature Conservation and Nuclear Safety, September 2005 [http://www.bmu.de/files/english/nuclear_safety/application/pdf/2nationaler_bericht_atomenergie_en. pdf]

Bodman, Samuel. "Statement From Secretary of Energy, Samuel Bodman," >*energy. gov*, 16 March 2005 [http://www.energy.gov/engine/content.do?PUBLIC_ID= 17629&BT_CODE= PR_PRESSRELEASES&TT_CODE= PRESSRELEASE]

Bodman, Samuel W. Letter to the Honorable Richard B. Cheney supporting the "Nuclear Fuel Management and Disposal Act." March 6, 2007 [http://www.ocrwm.doe.gov/info_library/newsroom/documents/2007bill.pdf]

Bryson, Amy Joi. "Utah Group just says no to N-waste at Goshute," *Deseret News*, 27 April 2002.

Bunn, Mathew, et. al. "Interim Storage of Spent Nuclear Fuel: A Safe, Flexible, and Cost Effective Near-Term Approach to Spent Fuel Management," A Joint Report from the Harvard University Project on Managing the Atom and the University of Tokyo Project on Sociotechnics of Nuclear Energy, June 2001 [http://ksgnotes1.harvard.edu/BCSIA/Library.nsf/pubs/spentfuel]

Bush, George W. "Presidential Letter to Congress," 15 February 2002 [http://www.whitehouse.gov/news/releases/2002/02/20020215-10.html]

Callan, Joseph L. "NRC's Report to Congress on the Price-Anderson Act," U.S. Nuclear Regulatory Commission, SECY-98-160, 2 July 1998 [http://www.nrc.

gov/reading-rm/doc-collections/commission/secys/1998/secy1998-160/1998-160scy.html]

Callan, Joseph L. "Risk-Informed, Performance-Based and Risk-Informed, Less-Prescriptive Regulation in the Office of Nuclear Material Safety and Safeguards," U.S. Nuclear Regulatory Commission, No. SECY-98-138, 11 June 1998 [http://www.nrc.gov/reading-rm/doc-collections/commission/secys/1998/secy1998-138/1998-138scy.html]

Carter, Luther J. *Nuclear Imperatives and Public Trust: Dealing with Radioactive Waste*, Washington D.C.: Resources for the Future.

Carter, Louis J. and Thomas H. Pigford "A Better Way at Yucca Mountain," Preliminary Report of Work in Progress, Meeting of the National Research Council's Board on Radioactive Waste Management, 20 September 2004.

CBS News, "EPA Revises Nuclear Dump Standards," 9 August 2005 [http://www.cbsnews.com/stories/2005/08/09/tech/main768203.shtml?CMP=ILC-SearchStories]

Clarke, Tracylee. "An Ideographic Analysis of Native American Sovereignty in the State of Utah: Enabling Denotative Dissonance and Constructing Irreconcilable Conflict," *Wicazo Sa Review*, University of Minnesota Press, vol. 17, No. 2, Fall 2002: 43-63.

CNR, *Canadian National Report for the Joint Convention on the Safety of Spent Fuel Management and on the Safety of Radioactive Waste Management, Second Report*, October 2005 [http://www.nuclearsafety.gc.ca/pubs_catalogue/uploads/2005_joint_convention_report_English.pdf]

Courier, "Court: U.S. must take nuclear waste by 1998," *The Courier*, 24 July 1996: A1.

Coyne, Connie. "Utah Site Foes Call Approval of Yucca Waste a Bitter Pill," *The Salt Lake Tribune*, 16 February 2002 [http://www.energy-net.org/IS/EN/NUKE/WST/FED/NEWS/NP-1644.202]

CRCPD, Conference of Radiation Control Program Directors [http://crcpd.org/Transportation_related_docs.asp]

CRCPD, "Directory of State and Federal Agencies Involved with the Transportation of Radioactive Material," Conference of Radiation Control Program Directors, CRCPD Publication E-04-6, October 2004 [http://crcpd.org/Transportation/TransDir04.pdf]

Defra, *The United Kingdom's Second National Report on Compliance with the Obligations of the Joint Convention on the Safety of Spent Fuel Management and on the Safety of Radioactive Waste Management*, Department for Environment, Food and Rural Affairs, February 2006 [http://www.defra.gov.uk/environment/radioactivity/government/international/pdf/jointconreport06.pdf]

DHS, "The National Strategy for the Physical Protection of Critical Infrastructure and Key Assets," U.S. Department of Homeland Security, 14 February 2003, [http://www.dhs.gov/xlibrary/assets/Physical_Strategy.pdf]

DOE, "About DOE: Mission," U.S. Department of Energy [http://www.doe.gov/engine/content.do?BT_CODE=ABOUTDOE]

DOE, "Damage to FOC from '92 quake less than first cited," *Of Mountains & Science*, U.S. Department of Energy, Yucca Mountain Project Studies, Winter 1996: 129-40.

DOE, *DOE's Yucca Mountain Studies*, Yucca Mountain Site Characterization Project, U.S. Department of Energy, Office of Civilian Radioactive Waste Management, DOE/RW-0345P, December 1992.

DOE, "DOE Researchers Demonstrate Feasibility of Efficient Hydrogen Production from Nuclear Energy," U.S. Department of Energy, Press Release, 30 November 2004 [http://www.energy.gov/news/1545.htm]

DOE, "DOE to Send Proposed Yucca Mountain Legislation to Congress," Office of Public Affairs, U.S. Department of Energy, March 6, 2007 [http://www.ocrwm. doe.gov/info_library/newsroom/documents/Yucca_leg_03_06_07_press_release. pdf]

DOE, "Draft Environmental Impact Statement for a Geologic Repository for the Disposal of Spent Nuclear Fuel and High-Level Radioactive Waste at Yucca Mountain, Nye County, Nevada," U.S. Department of Energy, Office of Civilian Radioactive Waste Management, vol. II, Appendix J, DOE/EIS-0250D, July 1999.

DOE, "Drillers find more perched water in Yucca Mountain's unsaturated zone," *Of Mountains & Science*, U.S. Department of Energy, Yucca Mountain Project Studies, Summer 1995 [http://www.ymp.gov/ref_shlf/ofms/default.html]

DOE, "Final Environmental Impact Statement for a Geologic Repository for the Disposal of Spent Nuclear Fuel and High-Level Radioactive Waste at Yucca Mountain, Nye County, Nevada: Readers Guide and Summary," U.S. Department of Energy, Office of Civilian Radioactive Waste Management, DOE/EIS-0250F, February 2002.

DOE, "First Agreement Reached with Utility on Nuclear Waste Acceptance," *Department of Energy News*, U.S. Department of Energy, 20 July 2000 [http:// www.energy.gov/HQPress/releases00/julpr/pr00186.htm]

DOE, "Nuclear Power 2010," Office of Nuclear Energy, Science and Technology, U.S. Department of Energy, updated: 8/27/04 [http://nuclear.gov/NucPwr2010/ NucPwr2010.html]

DOE, "Nuclear Power 2010," U.S. Department of Energy, Office of Nuclear Energy, Science and Technology, updated: September 2006 [http://np2010.doc.html]

DOE, "Project scientists assess tritium traces for evidence of fast routes to repository," *Of Mountains & Science*, U.S. Department of Energy, Yucca Mountain Project Studies, Winter 1996: 134, 139.

DOE, "Radiological Assistance Program," U.S. Department of Energy, Office of Defense Programs, Order DOE 5530.3, 14 January 1992 [http://www.directives. doe.gov/pdfs/doe/doetext/oldord/5530/o55303c1.pdf]

DOE, "Site Characterization Progress Report: Yucca Mountain, Nevada," U.S. Department of Energy, DOE/RW-0496, No. 15, April 1997.

DOE, "Spent Nuclear Fuel Transportation," U.S. Department of Energy, Office of Public Affairs [http://www.ocrwm.doe.gov/transport/pdf/snf_trans.pdf]

DOE, "Technical Report Confirms Reliability of Yucca Mountain Technical Work," U.S. Department of Energy, 17 February 2006 [http://www.energy.gov/news/3220. htm]

DOE, "Transporting Radioactive Materials: Answers to Your Questions," U.S. Department of Energy, Office of Civilian Radioactive Waste Management, National Transportation Program, June 1999 [http://www.ocrwm.doe.gov/wat/ transportation.shtml]

DOE, "Transporting Spent Nuclear Fuel and High-Level Radioactive Waste to a National Repository: Answers to Frequently Asked Questions," U.S. Department of Energy, Office of Civilian Radioactive Waste Management, Yucca Mountain Project, December 2002 [http://www.ocrwm.doe.gov/wat/pdf/snf_transfaqs.pdf]

DOE, "Yucca Mountain Volcanic Hazard Analysis Completed," Yucca Mountain Project Studies, U.S. Department of Energy, 24 September 1996, [http://www. ymp.gov/wha_news/ymppr/pvha.htm]

DOI, "Record of Decision for the Construction and Operation of an Independent Spent Fuel Storage Installation (ISFSI) on the Reservation of the Skull Valley Band of Goshute Indians (band) in Tooele County, Utah," U.S. Department of Interior, Bureau of Indian Affairs, 7 September 2006 [http://www.deq.utah.gov/ Issues/no_high_level_waste//ROD%20PFS%2009072006.pdf]

DOI, "Record of Decision Addressing Right-of-Way Applications U 76985 and U 76986 to Transport Spent Nuclear Fuel To the Reservation of the Skull Valley Band of Goshute Indians," U.S. Department of Interior, Bureau of Land Management, 7 September 2006 [http://www.deq.utah.gov/Issues/no_high_level_waste/ROD%2 0Right%20of%20Way%20Skull%20Valley%2009072006.pdf]

EIA, "World Cumulative Spent Fuel Projections by Region and Country," Energy Information Agency, U.S. Department of Energy, 1 May 2001 [http://www.eia. doe.gov/cneaf/nuclear/page/forecast/cumfuel.html]

EIA, "Annual Energy Outlook 2004," Energy Information Administration, U.S. Department of Energy, Table A8 (Electric Supply, Disposition, Prices and Emissions): 145 [http://www.eia.doe.gov/oiaf/archive/aeo04/pdf/0383(2004). pdf]

EM, "Environmental Management...Making Accelerated Cleanup a Reality," U.S. Department of Energy, Office of Environmental Management [http://www. em.doe.gov]

ENR, "High Court Leaves DOE on the Hook," *Engineering News Record*, 30 November/7 December 1998.

EOP, "Statement of Administration Policy: S. 104 – Nuclear Waste Policy Act of 1997," Executive Office of the President, 7 April 1997 [http://clinton3.nara.gov/ OMB/legislative/sap/105-1/S104-s.html]

EPA, "Radiation Protection at EPA: The First 30 Years...Protecting People and the Environment," Office of Radiation and Indoor Air, U.S. Environmental Protection Agency, EPA 402-B-00-001, August 2000 [http://www.epa.gov/radiation/docs/ 402-b-00-001.pdf]

EPA, "About EPA's Radiation Protection Program," Radiation Information, U.S. Environmental Protection Agency [http://www.epa.gov/radiation/about/index. html]

EPA, "Superfund Risk Assessment," U.S. Environmental Protection Agency, Waste and Cleanup Risk Assessment [http://www.epa.gov/oswer/riskassessment/risk_superfund.htm]

Erickson, Kai. "Out of Sight, Out of Our Minds," *The New York Times Magazine*, 6 March 1994: 36-41, 50, 63.

FAEA, *The National Report of the Russian Federation on Compliance with the Obligations of the Joint Convention on the Safety of Spent Fuel Management and the Safety of Radioactive Waste Management*, Prepared for the Second Review Meeting in Frames of the Joint Convention on the Safety of Spent Fuel Management and the Safety of Radioactive Waste Management, Federal Atomic Energy Agency, Moscow 2006 [http://www-ns.iaea.org/downloads/rw/conventions/russian-federation-national-report.pdf]

Fahys, Judy. "Feds Are No Help to Utah on N-Waste," *The Salt Lake Tribune*, 19 December 2002 [http://www.sltrib.com/2002/dec/12192002/Utah/12796.asp]

Fahys, Judy. "Goshute plan suffers setback," *The Salt Lake Tribune*, 27 September 2003 [http://www.sltribune.com/2003/sep/09272003/utah/96424.asp]

Fahys, Judy. "Senator Cites N-Dump Politics," *The Salt Lake Tribune*, 27 April 2002 [http://www.energy-net.org/IS/EN/NUKE/WST/FED/NEWS/NP-2710.402]

Fahys, Judy. "State Files Its Appeal to Halt Goshute Plan," *The Salt Lake Tribune*, 16 August 2002 [http://www.energy-net.org/IS/EN/NUKE/WST/FED/NEWS/NP-1627.802]

Fahys, Judy. "Utah Fears Waste Plan Is Shoo-In," *The Salt Lake Tribune*, 22 April 2002 [http://www.energy-net.org/IS/EN/NUKE/WST/FED/NEWS/NP-2242.402]

Fahys, Judy and Dan Harrie. "'Plan B' Aims to Outbid Goshutes' N-Waste Site,'" *The Salt Lake Tribune*, 6 February 2003 [http://www.sltrib.com/2003/feb/02062003/Utah/26765.asp]

Falcone, Santa and Kenneth Orosco. "Coming Through a City Near You: The Transport of Hazardous Wastes," *Policy Studies Journal*, vol. 26, Issue 4, Winter 1998: 760-73.

FANC, *Second Meeting of the Contracting Parties to the Joint Convention on the Safety of Spent Fuel Management and on the Safety of Radioactive Waste Management*, Federal Agency for Nuclear Control, Kingdom of Belgium, National Report, May 2006 [http://www.avnuclear.be/avn/JointConv2006.pdf]

Federal Register, "Office of Civilian Radioactive Waste Management; Safe Routine Transportation and Emergency Response Training; Technical Assistance and Funding," vol. 63, No. 83, 30 April 1998.

FEMA, "Backgrounder: Nuclear Power Plant Emergency," Hazards, Federal Emergency Management Agency [http://www.fema.gov/hazards/nuclear/radiolo.shtm]

FEMA, "Guidance for Developing State and Local Radiological Emergency Response Plans and Preparedness for Transportation Accidents," Federal Emergency Management Agency, FEMA-REP-5, Draft Revision, August 1988.

Flint, Lawrence. "Shaping Nuclear Waste Policy at the Juncture of Federal and State Law," *Boston College Environmental Affairs Law Review*, vol. 28, No. 1, 2000

[http://www.bc.edu/schools/law/lawreviews/meta-elements/journals/bcealr/28_1/05_FMS.htm]

Flynn, James H., C.K. Mertz, and Paul Slovic. "Results of a 1997 National Nuclear Waste Transportation Survey", Decision Research, January 1998 [http://www.state.nv.us/nucwaste/trans/1dr01a.htm]

FMCSA, "About Us," U.S. Department of Transportation, Federal Motor Carrier Safety Administration [http://www.fmcsa.dot.gov/aboutus/aboutus.htm]

FRA, "Hazardous Materials Division," Federal Railroad Administration, U.S. Department of Transportation [http://www.fra.dot.gov/us/content/337]

FRA, "Safety Compliance Oversight Plan for Rain Transportation of High-Level Radioactive Waste & Spent Nuclear Fuel," U.S. Department of Transportation, Federal Railroad Administration, June 1998, [http://www.fra.dot.gov/downloads/safety/scopfnl.pdf]

Freking, Kevin. "Nuclear plant neighbors worried about waste," *Arkansas Democrat Gazette*, 30 April 1995, p. 1A, col. 1.

FRERP, "Federal Radiological Emergency Response Plan," May 1, 1996. Summary available in the *Federal Register*, vol. 61, No. 90, Wednesday, 8 May 1996, Notice: 20 944.

GAO, "Nuclear Waste: Technical, Schedule, and Cost Uncertainties of the Yucca Mountain Repository Project," U.S. Government Accountability Office, GAO-02-191, December 2001 [http://www.gao.gov/new.items/d02191.pdf]

GAO, "Spent Nuclear Fuel: Options Exist to Further Enhance Security," U.S. Government Accountability Office, Report to the Chairman, Subcommittee on Energy and Air Quality, Committee on Energy and Commerce, U.S. House of Representatives, GAO-03-426, July 2003 [http://www.gao.gov/new.items/d03426.pdf]

GAO, "Nuclear Regulatory Commission: Preliminary Observations on Efforts to Improve Security at Nuclear Power Plants," U.S. Government Accountability Office, Statement of Jim Wells (Director, National Resources and Environment), Testimony Before the Subcommittee on National Security, Emerging Threats, and International Relations, Committee on Government Reform, House of Representatives, GAO-04-106T, 14 September 2004 [http://www.gao.gov/new.items/do041064t.pdf]

Geringer, Jim. "Geringer Says No to Owl Creek Nuclear Waste Project," Office of the Governor, State of Wyoming, 6 April 1998.

Gowda, M.V. Rajeev and Doug Easterling. "Nuclear Waste and Native America: The MRS Siting Exercise," *Risk Health, Safety and Environment*, Summer 1998: 229-58.

GPO, "Budget of the United States Government: Fiscal Year 2008," U.S. Government Printing Office, 2007 [http://www.gpoaccess.gov/usbudget/fy08/pdf/budget/energy.pdf]

Guinn, Kenny C. "Official Notice of Disapproval of the Yucca Mountain Site," State of Nevada, Office of the Governor, 8 April 2002.

Hall, Jim. "Testimony of Jim Hall on Behalf of the Transportation Safety Coalition," U.S. Senate, Committee on Energy and Natural Resources, 23 May 2002 [http://www.yuccamountain.org/pdf/hall052302.pdf]

Halstead, Robert J. "Radiation Exposures From Spent Nuclear Fuel and High-Level Nuclear Waste Transportation to a Geologic Repository or Interim Storage Facility in Nevada," State of Nevada, Nuclear Waste Project Office, no date [http://www.state.nv.us/nucwaste/trans/radexp.htm]

Hebert, H. Josef. "Nuclear waste could be terror target," *The Courier*, 20 August 2002, p. 6, col. 1.

Hebert, H. Josef. "Senate backs plan to revive nuclear power," *The Courier*, 11 June 2003, p. 9, col. 1.

Highfield, Roger. "Seabed scars show power of ocean quake," *The UK Telegraph On-Line*, 10 February 2005 [http://telegraph.co.uk/news/main.jhtml;sessionid= FZFDIIAMDUEKDQFIQMFCM5OAVCBQYJVC?xml=/news/2005/02/10/ wsea10.xml&secureRefresh=true&_requestid=59553]

Holt, Mark. "Civilian Nuclear Waste Disposal," Congressional Research Service, RL33461 (Updated 19 September 2006) [http:ncseonline.org/NLE/CRSreports/ 06Sep/RL33461.pdf]

Holt, Mark. "Civilian Nuclear Spent Fuel Temporary Storage Options," U.S. Congressional Research Service, 96-212 ENR (Updated 27 March 1998) [http:// cnie.org/NLE/CRSreports/Waste/waste-20.cfm]

Holt, Mark. "Civilian Nuclear Waste Disposal," *CRS Issue Brief*, U.S. Congressional Research Service, 21 November 1996, [http://www.cnie.org/nle/waste-2.html]

Holt, Mark. "Transportation of Spent Nuclear Fuel," Congressional Research Service, CRS Report for Congress, 97-403 ENR (Updated 29 May 1998) [http:// cnie.org/NLE/CRSreports/energy/eng-34.cfm]

Holt, Mark and Zachary Davis, "Nuclear Energy Policy," *CRS Issue Brief*, U.S. Congressional Research Service (Updated 5 December 1996) [http://www.cnie. org/nle/eng-5.html]

HSK, *Implementation of the Obligations of the Joint Convention on the Safety of Spent Fuel Management and on the Safety of Radioactive Waste Management*, Second National Report of Switzerland in Accordance with Article 32 of the Convention, Department of Environment, Transport, Energy and Communications, September 2005 [http://www.hsk.ch/english/files/pdf/joint-convention_CH_sept-05.pdf]

IAEA, *Nuclear Power and Sustainable Development*, International Atomic Energy Agency, April 2006 [http://www.iaea.org/Publications/Booklets/Development/ npsd0506.pdf]

IAEA, *Nuclear Technology Review 2006*, International Atomic Energy Agency, IAEA/NTR/2006, August 2006 [http://www.iaea.org/OurWork/ST/NE/Pess/ assets/ntr2006.pdf]

IAEA, "The Long Term Storage of Radioactive Waste: Safety and Sustainability: A Position of International Experts," International Atomic Energy Agency, June 2003 [http://www-pub.iaea.org/MT/publications/PDF/LTS-RW_web.pdf]

IAEA, "Nuclear Power and Sustainable Development," International Atomic and Energy Agency, April 2006 [http://www.iaea.org/Publications/Booklets/ Development/npsd0506.pdf]

IEER, "International Repository Programs," Institute for Energy and Environmental Research, May 1999 [http://ieer.org/sdafiles/vol_7/7-3/repos.html]

Jackson, Shirley M. "Transitioning to Risk-Informed Regulation: The Role of Research," U.S. Nuclear Regulatory Commission, Office of Public Affairs, 26th Annual Water Reactor Safety Meeting, 26 October 1998, No. S-98-26 [http://www. nrc.gov/reading-rm/doc-collections/commission/speeches/1998/s98-26.html]

Jacob, Gerald. *Site Unseen: The Politics of Siting a Nuclear Waste Repository*, Pittsburgh, PA: University of Pittsburgh Press, 1990.

Jones, Tom A. Former lobbyist for Owl Creek Energy Project, Telephone Interview, 29 March 2005.

Jordan, "S.C. to block plutonium trucks," *Arkansas Democrat Gazette*, 15 June 2002, p. 2A, col. 5.

Katz, Jonathan L. "A Web of Interests: Stalemate on the Disposal of Spent Nuclear Fuel," Policy Studies Journal, vol. 29, Issue 3, 2001: 456-77.

Kraft, Michael E. "Democratic Dialogue and Acceptable Risks: The Politics of High-Level Nuclear Waste Disposal in the United States," in *Siting by Choice: Waste Facilities, NIMBY, and Volunteer Communities*, Don Munton, ed., (Georgetown University Press: Washington D.C., 1996): 108-41.

Kugler, *Policy Analysis in National Security Affairs: New Methods for a New Era*, Center for Technology and Security Policy, National Defense University (NDU Press: Washington D.C., 2006).

Lamb, Matthew, Marvin Resnikoff and Richard Moore. "Worst Case Credible Nuclear Transportation Accidents: Analysis for Urban and Rural Nevada," Radioactive Waste Management Associates, August 2001 [http://www.state. nv.us/nucwaste/trans/rwma0108.pdf]

London Convention on the Prevention of Marine Pollution by Dumping of Wastes and Other Matter, 1972, Annexes I and II, amended in 1993. [http://www. londonconvention.org]

Loux, Robert R. "Nuclear Waste Policy at the Crossroads," [http://www.astro.com/ yucca]

Lowry, David. "2010: America's Nuclear Waste Odyssey," *New Scientist*, 6 March 1993: 30-33.

LVRJ, "E-mails say scientists fabricated quality assurance on Yucca Mountain," *Las Vegas Review Journal*, 2 April 2005 [http:www.reviewjournal.com/lvrj_home/ 2005/Apr-02-Sat-2005/news/26204008.html]

LVRJ, "Energy officials turn shy in talk about Yucca Schedule," *Las Vegas Review Journal*, 11 March 2005 [http:www.reviewjournal.com/lvrj_home/2005/Mar-11-Fri-2005/news/26048040.html]

LVRJ, "Yucca Mountain: Guinn vetoes Bush," *Las Vegas Review-Journal*, 9 April 2002 [http:www.lvrj.com/cgi-bin/printable.cgi?/lvrj_home/2002/Apr-09-Tue-2002/news/18476447]

Macfarlane, Alison. "Interim Storage of Spent Fuel in the United States," *Annual Review of Energy and the Environment*, vol. 26, 2001: 201-35.

Macfarlane, Allison. "The Problem of Used Nuclear Fuel: Lessons for Interim Solutions From a Comparative Cost Analysis," *Energy Policy*, vol. 29, 2001: 1379-89.

Martin, Sue. Spokesperson for Private Fuel Storage LLC, Telephone Interview, 22 March 2002.

Martin, Sue. Spokesperson for Private Fuel Storage LLC, Telephone Interview, 5 September 2002.

Martin, Sue. Spokesperson for Private Fuel Storage LLC, Telephone Interview, 6 February 2003.

Martin, Sue. Spokesperson for Private Fuel Storage LLC, Telephone Interview, 8 October 2003.

Martin, Sue. Spokesperson for Private Fuel Storage LLC, Telephone Interview, 23 January 2004.

McCutcheon, Chuck. *Nuclear Reactions: The Politics of Opening a Radioactive Waste Disposal Site*, University of New Mexico Press, 2002.

Meserve, Richard A. Letter to Congressman Edward J. Markey, 4 March 2002 [http://www.house.gov/markey/ISS_nuclear_ltr020325a.pdf]

Meserve, Richard A. "Nuclear Power Plant Security," Statement submitted by the NRC to the Committee on Environment and Public Works, U.S. Senate, 5 June 2002 [http://www.nrc.gov/reading-rm/doc-collections/congress-docs/congress-testimony/2002/ml021570116.pdf]

METI, *Joint Convention on the Safety of Spent Fuel Management and on the Safety of Radioactive Waste Management*, National Report of Japan for the Second Review Meeting, October 2005 [http://www.meti.go.jp/english/report/index.html]

MITYC, *Joint Convention on the Safety of Spent Fuel Management and on the Safety of Radioactive Waste Management*, Second Spanish National Report, October 2005 [http://www.mityc.es/NR/rdonlyres/BF3E47F5-7861-4A28-8FB3-8F2DBAB183A3/0/ConvencionInforme2_ing.pdf]

Moore, Robert. "Public Service, Personal Gain in Wyoming," Center for Public Integrity, 21 May 2000 [http://www.50statesonline.org/cgi-bin/50states/states.asp?State=WY&Display=StateSummary]

MSDS, *Sweden's Second National Report Under the Joint Convention on the Safety of Spent Fuel Management and on the Safety of Radioactive Waste Management*, Ministry of Sustainable Development, Ds 2005:44, 2005 [http://www.sweden.gov.se/content/1/c6/05/40/89/fc570cf2.pdf]

Munton, Don. "Introduction: The NIMBY Phenomenon and Approaches to Facility Siting," in *Hazardous Waste Siting and Democratic Choice: The NIMBY Phenomenon and Approaches to Facility Siting*, Georgetown University Press, 1996.

NAC, "Energy Solutions, Information and Technology," NAC International, 2003 [http://www.nacintl.com]

NACO, "Environment, Energy & Land Use," *The American County Platform and Resolutions*, Resolution 3A, Adopted July 16, 2002 [http://www.naco.org/Content/ContentGroups/Legislative_Affairs/Advocacy1/EELU/EELU1/3EELU_02-03.pdf]

NARUC, "Resolution Regarding Guiding Principles for Disposal of High-Level Nuclear Waste," National Association of Regulatory Utility Commissioners, 2000 Resolutions and Policy Positions (Electricity) [http://www.naruc.org/Resolutions/2000/annual/elec/disposal_nuclear_waste.shtml]

NARUC, "Resolution Supporting Reform of the Nuclear Waste Fund," National Association of Regulatory Utility Commissioners, 2004 Resolutions and

Policy Positions (Electricity) [http://www.naruc.org/displaycommon.cfm?an= 1&subarticlenbr=295]

NARUC, "Resolution Supporting Reform of the Nuclear Waste Fund," National Association of Regulatory Utility Commissioners, 2005 Resolutions and Policy Positions: Electricity [http://www.naruc.org/displaycommon.cfm?an= 1&subarticlenbr=394]

NAS, "Building Consensus Through Risk Assessment and Management of the Department of Energy's Environmental Remediation Program," Committee to Review Risk Management in the DOE's Environmental Remediation Program, National Academy of Sciences, National Research Council, National Academy Press, 1994 [http://www.nap.edu/openbook/NI000191/html/R1.html]

NAS, "Developing 'Risk-Based' Approaches for Disposition of Transuranic and High-Level Radioactive Waste," National Academy of Sciences, Current Projects, BRWM-U-03-01-A [http://www.nas.edu]   The published report, *Risk and Decisions About Disposition of Transuranic and High-Level Radioactive Waste*) is available at [http://newton.nap.edu/catalog/11223.html#toc]

NAS, "Disposition of High-Level Waste and Spent Nuclear Fuel: The Continuing Societal and Technical Challenges," Committee on Disposition of High-Level Radioactive Waste Through Geological Isolation, Board on Radioactive Waste Management, Division on Earth and Life Studies, National Academy of Sciences, National Research Council, National Academy Press, Washington D.C., 2001 [http://www.nap.edu/openbook/0309073170/html/R1.html]

NAS, "Safety and Security of Commercial Spent Nuclear Fuel Storage," Committee on the Safety and Security of Commercial Spent Nuclear Fuel Storage, Board on Radioactive Waste Management, Division on Earth and Life Studies, National Academy of Sciences, National Research Council, National Academy Press, 1994 [http://www.nap.edu/openbook/0309096472.html]

NAS, "The Disposal of Radioactive Waste On Land," Report of the Committee on Waste Disposal of the Division of Earth Sciences, National Academy of Sciences, National Research Council, Publication 519, September 1957 [http://www.nap.edu/books/NI000379/html]

NCEP, "Ending the Energy Stalemate: A Bipartisan Strategy to Meet America's Energy Challenges, The National Commission on Energy Policy, December 2004 [http://64.70.252.93/newfiles/Final_Report/index.pdf]

NCSL, "Radioactive Materials Transportation: Update for the High-Level Radioactive Waste Working Group," National Conference of State Legislatures, Environment, Energy and Transportation Program, September 1998 [http://www.ncsl.org/programs/esnr/radmatup.htm]

NEA, *Safety of Geological Disposal of High-level and Long-lived Radioactive Waste in France*, An International Peer Review of the "Dossier 2005 Argile" Concerning Disposal in the Callovo-Oxfordian Formation, Nuclear Energy Agency, Organisation for Economic Co-operation and Development, NEA No. 6178, OECD 2006 [http://www.nea.fr/html/rwm/reports/2006/nea6178-argile.pdf]

NEI, "Appellate Court Ruling Likely Makes Taxpayers Liable for DOE Failure to Dispose of Used Nuclear Fuel," News, Nuclear Energy Institute, 2002 [http://www.nei.org/doc.asp?Print=true&DocID=980&CatNum=&CatID=]

NEI, *Industry Spent Fuel Storage Handbook*, Nuclear Energy Institute, NEI 98-01, May 1998.

NEI, "Post-Sept. 11 Security Enhancements: More Personnel, Patrols, Equipment, Barriers," Nuclear Energy Institute, Safety and Security report [http://www.nei.org/index.asp?catnum= 2&catid=275]

NEI, "Powering Tomorrow…With Clean, Safe Energy," Vision 2020, Nuclear Energy Institute [http://www.nei.org/documents/Vision2020_Backgrounder_Powering_Tomorrow.pdf]

NEI, "Safe Interim On-Site Storage of Used Nuclear Fuel," Nuclear Energy Institute, Nuclear Waste Disposal report [http://www.nei.org/index.asp?catnum=2&catid=70]

NEI, "Used Nuclear Fuel Management," Nuclear Energy Institute, October 2004 [http://www.nei.org/doc.asp?catnum=3&catid=300&docid=&format=print]

Nevada, "Reported Incidents Involving Spent Nuclear Fuel Shipments 1949 to Present," State of Nevada, 6 May 1996 [http://www.state.nv.us/nucwaste/trans/nucinc01.htm]

Nixon, Will. "High Energy," *E/The Environmental Magazine*, May/June 1995, [http://www.adams.ind.net/Environment.html]

NO Coalition, "White Paper Regarding Opposition to the High-Level Nuclear Waste Storage Facility Proposed By Private Fuel Storage On The Skull Valley Band of Goshute Indian Reservation, Skull Valley, Utah," The Coalition Opposed to High-Level Nuclear Waste, 28 November 2000.

NRC, *Disposition of High-Level Waste and Spent Nuclear Fuel: The Continuing Societal and Technical Challenges*, U.S. Nuclear Regulatory Commission, Committee on Disposition of High-Level Radioactive Waste Through Geological Isolation, Board on Radioactive Waste Management, Division on Earth and Life Sciences (National Academies Press: Washington D.C., 2001)

NRC, "Dry Cask Storage of Spent Nuclear Fuel," Backgrounder, Office of Public Affairs, U.S. Nuclear Regulatory Commission, December 2004 [http://www. nrc. gov/reading-rm/doc-collections/fact-sheets/dry-cask-storage.html]

NRC, "Emergency Preparedness and Response," U.S. Nuclear Regulatory Commission [http://www.nrc.gov/what-we-do/emerg-preparedness.html]

NRC, "Final Environmental Impact Statement for the Construction and Operation of an Independent Spent Fuel Storage Installation on the Reservation of the Skull Valley Band of Goshute Indians and the Related Transportation Facility in Tooele County, Utah," U.S. Nuclear Regulatory Commission, Office of Nuclear Material Safety and Safeguards, NUREG-1714, vol. 1, December 2001 [http://www.nrc. gov/]

NRC, "High-Level Waste," U.S. Nuclear Regulatory Commission, Revised June 23, 2003 [http://www.nrc.gov/waste/high-level-waste.html]

NRC, "Locations of Independent Spent Fuel Storage Installations," U.S. Nuclear Regulatory Commission, Revised September 25, 2003 [http://www.nrc.gov/waste/spent-fuel-storage/locations.html]

NRC, "Memorandum and Order," U.S. Nuclear Regulatory Commission, Docket No. 72-22-ISFSI, CLI-05-19, 9 September 2005 [http://www.nrc.gov/reading-rm/doc-collections/commission/orders/2005/2005-19cli.pdf]

NRC, "NRC Commissioner Jaczko Takes Oath of Office; Commissioner Lyons Swearing-In Set for Next Week," *NRC News*, U.S. Nuclear Regulatory Commission, No. 05-013, 21 January 2005 [http://www.nrc.gov/reading-rm/doc-collections/news/2005/05-013.pdf]

NRC, "NRC Denies Utah's Final Appeals, Authorizes Staff to Issue License for PFS Facility," *NRC News*, U.S. Nuclear Regulatory Commission, No. 05-126, September, 2005 [http://www.nrc.gov/reading-rm/doc-collections/news/2005/05-126.html]

NRC, "NRC Increases Secondary Premium for Nuclear Power Plant Insurance," *NRC News*, U.S. Nuclear Regulatory Commission, 4 August 2003 [http://www.nrc.gov/reading-rm/doc-collections/news/2003/03-100.html]

NRC, "NRC Licenses Spent Nuclear Fuel Storage Facility at Idaho National Engineering and Environmental," *NRC News*, U.S. Nuclear Regulatory Commission, No. 04-150, 1 December 2004 [http://www.nrc.gov/reading-rm/doc-collections/news/2004/04-150.htm]

NRC, "NRC Publishes Final Environmental Impact Statement on Proposed Spent Fuel Storage Facility," *NRC News*, U.S. Nuclear Regulatory Commission, No.02-002 [http://www.nrc.gov/reading-rm/doc-collections/news/archive/02-002.html]

NRC, "Nuclear Fuel Pool Capacity," U.S. Nuclear Regulatory Commission, Revised 23 June 2003 [http://www.nrc.gov/waste/spent-fuel-storage/nuc-fuel-pool.html]

NRC, "Nuclear Materials," U.S. Nuclear Regulatory Commission [http://www.nrc.gov/materials.html]

NRC, "Nuclear Reactors," U.S. Nuclear Regulatory Commission [http://www.nrc.gov/reactors.html]

NRC, "Nuclear Security–Before and After September 11," U.S. Nuclear Regulatory Commission [http://www.nrc.gov/what-we-do/safeguards/response-911.html]

NRC, "Radioactive Waste," U.S. Nuclear Regulatory Commission [http://www.nrc.gov/waste.html]

NRC, "Safety and Spent Fuel Transportation," U.S. Nuclear Regulatory Commission, NUREG/BR-0292, March 2003 [http://www.nrc.gov/reading-rm/doc-collections/nuregs/brochures/br0292]

NRC, "Spent Fuel Pools," U.S. Nuclear Regulatory Commission, Revised 23 June 2003 [http://www.nrc.gov/waste/spent-fuel-storage/pools.html]

NRC, "Strategic Assessment Issue: 6. High-Level Waste and Spent Fuel," U.S. Nuclear Regulatory Commission, 16 September 1996 [http://www.nrc.gov/NRC/STRATEGY/ISSUES/dsi06isp.htm]

NRC, "Transporting Spent Nuclear Fuel," Diagram and Description of Typical Spent Fuel Transportation Cask, U.S. Nuclear Regulatory Commission [http://www.nrc.gov/waste/spent-fuel-storage/diagram-typical-trans-cask-system.doc]

NRC, "What We Do," U.S. Nuclear Regulatory Commission [http://www.nrc.gov/what-we-do.html]

NRC, "Who We Are," U.S. Nuclear Regulatory Commission [http://www.nrc.gov/who-we-are.html]

NSC, "Backgrounder on the Waste Isolation Pilot Plant: How Are Transportation Routes Chosen?" National Safety Council, Environmental Health Center, August 2001 [http://www.nsc.org/public/ehc/rad/backgr8.pdf]

NTP, "National Transportation Program Products & Services," National Transportation Program, U.S. Department of Energy [http://www.ntp.doe.gov/mission.html]

NTP, "Questions and Answers About the Program," National Transportation Program, U.S. Department of Energy, November 2000 [http://www.ntp.doe.gov/qa.html]

Nuke-Energy, "Federal Judge Rules Against Utah Nuclear Waste Laws," [http://www.nuke-energy.com/data/stories/aug02/fed%20judge.htm]

Nuke-Energy, "OCRWM Chief Says Something: 2010 Milestone is 'Tight'," Nuke-Energy, July 2002 [http://www.nuke-energy.com/data/stories/July02/Chu.htm]

NWPA, Nuclear Waste Policy Act of 1982 (P.L. 97-425) [http://www.ocrwm.doe.gov/documents/nwpa/css/nwpa.htm]

NWTRB, *Disposal and Storage of Spent Nuclear Fuel–Finding the Right Balance*, Nuclear Waste Technical Review Board, March 1996 [http://www.nwtrb.gov/reports/storage.pdf]

NWTRB, "Nuclear Waste Management in the United States," Nuclear Waste Technical Review Board, Topseal Conference, June 1996 [http://www.nwtrb.gov/reports/wastemgt.pdf]

NWTRB, "What is the U.S. Nuclear Waste Technical Review Board?" U.S. Nuclear Waste Technical Review Board, NWTRB Viewpoint, November 2004 [http://www.nwtrb.gov/mission/nwtrb.pdf]

OCEP, "Owl Creek Energy Project: Private Interim Spent Fuel Storage Facility," Owl Creek Energy Project, Global Spent Fuel Management Summit, 3-6 December 2000.

OCRWM, "Annual Report to Congress–December 2004," Office of Civilian Radioactive Waste Management, U.S. Department of Energy, DOE/RW-0569 [http://www.ocrwm.doe.gov/pm/program_docs/annualreports/04ar/fy_2004.pdf]

OCRWM, "Appropriations by Fiscal Year [Yucca Mountain]," Office of Civilian Radioactive Waste Management Program, U.S. Department of Energy, [http://www.ocrwm.doe.gov/pm/budget/budget.shtml]

OCRWM, "Assessing the Future Safety of a Repository at Yucca Mountain," Office of Civilian Radioactive Waste Management, U.S. Department of Energy, CD-ROM, 1 March 2004.

OCRWM, "Belgium's Radioactive Waste Management Program," Office of Civilian Radioactive Waste Management, U.S. Department of Energy, DOE/YMP-0407, June 2001 [http://www.ocrwm.doe.gov/factsheets/doeymp0407.shtml]

OCRWM, "Canada's Radioactive Waste Management Program," Office of Civilian Radioactive Waste Management, U.S. Department of Energy, DOE/YMP-0408, June 2001 [http://www.ocrwm.doe.gov/factsheets/doeymp0408.shtml]

OCRWM, "China's Radioactive Waste Management Program," Office of Civilian Radioactive Waste Management, U.S. Department of Energy, DOE/YMP-0409, June 2001 [http://www.ocrwm.doe.gov/factsheets/doeymp0409.shtml]

OCRWM, "Civilian Radioactive Waste Management Program Plan (Revision 3)," Office of Civilian Radioactive Waste Management, U.S. Department of Energy, February 2000 [http://www.ocrwm.doe.gov/pm/pdf/pprev3.pdf]

OCRWM, "Cooperative Agreements," Office of Civilian Radioactive Waste Management, U.S. Department of Energy [http://ocrwm.doe.gov/wat/cooperative_agreements.shtml]

OCRWM, "Department of Energy Policy Statement for Use of Dedicated Trains for Waste Shipments to Yucca Mountain," Office of Civilian Radioactive Waste Management, U.S. Department of Energy, 2005 [http://www.ocrwm.doe.gov/transport/pdf/dts_policy.pdf]

OCRWM, "Disposal Options: Reprocessing and Transmutation," Office of Civilian Radioactive Waste Management, U.S. Department of Energy [http://ocrwm.doe.gov/ymp/about/reprocess.shtml]

OCRWM, "Final Environmental Impact Statement for a Geologic Repository for the Disposal of Spent Nuclear Fuel and High-Level Radioactive Waste at Yucca Mountain, Nye County, Nevada," Office of Civilian Radioactive Waste Management, U.S. Department of Energy, DOE/EIS-0250, February 2002 [http://www.ocrwm.doe.gov/documents/feis_2/vol_1/ch_06/index1_6.htm]

OCRWM, "Finland's Radioactive Waste Management Program," Office of Civilian Radioactive Waste Management, U.S. Department of Energy, DOE/YMP-0410, June 2001, available at: [http://www.ocrwm.doe.gov/factsheets/doeymp0410.shtml]

OCRWM, "France's Radioactive Waste Management Program," Office of Civilian Radioactive Waste Management, U.S. Department of Energy, DOE/YMP-0411, June 2001 [http://www.ocrwm.doe.gov/factsheets/doeymp0411.shtml]

OCRWM, "FY 2006 Budget Request Summary [Yucca Mountain]," Office of Civilian Radioactive Waste Management Program, U.S. Department of Energy, [http://www.ocrwm.doe.gov/pm/budget/budgetrollout_06/2006cbr7.shtml]

OCRWM, "Japan's Radioactive Waste Management Program," Office of Civilian Radioactive Waste Management, U.S. Department of Energy, DOE/YMP-0413, June 2001 [http://www.ocrwm.doe.gov/factsheets/doeymp0413.shtml]

OCRWM, "OCRWM Program Briefing," Office of Civilian Radioactive Waste Management, U.S. Department of Energy [http://www.ocrwm.doe.gov/pm/programbrief/briefing.htm]

OCRWM, "Oklo: Natural Nuclear Reactors," Office of Civilian Radioactive Waste Management, U.S. Department of Energy, DOE/YMP-0010, November 2004 [http://www.ocrwm.doe.gov/factsheets/doeymp0010.shtml]

OCRWM, "Overview," Office of Civilian Radioactive Waste Management, U.S. Department of Energy [http://www.ocrwm.doe.gov/overview.shtml]

OCRWM, "Radioactive waste: an international concern," Office of Civilian Radioactive Waste Management, U.S. Department of Energy, DOE/YMP-0405, June 2001 [http://www.ocrwm.doe.gov/factsheets/doeymp0405.shtml]

OCRWM, "Russia's Radioactive Waste Management Program," Office of Civilian Radioactive Waste Management, U.S. Department of Energy, DOE/YMP-0413, June 2001 [http://www.ocrwm.doe.gov/factsheets/doeymp0413.shtml]

OCRWM, "Science, Society and America's Nuclear Waste," Background Notes, Office of Civilian Radioactive Waste Management, U.S. Department of Energy [http://www.ocrwm.doe.gov/pm/program_docs/curriculum/unit_1_toc/14a.pdf]

OCRWM, "Spain's Radioactive Waste Management Program," Office of Civilian Radioactive Waste Management, U.S. Department of Energy, DOE/YMP-0415, June 2001 [http://www.ocrwm.doe.gov/factsheets/doeymp0415.shtml]

OCRWM, "Spent Nuclear Fuel Transportation," Office of Civilian Radioactive Waste Management, U.S. Department of Energy [http://www.ocrwm.doe.gov/wat/pdf/snf_trans.pdf]

OCRWM, "Sweden's Radioactive Waste Management Program," Office of Civilian Radioactive Waste Management, U.S. Department of Energy, DOE/YMP-0416, June 2001 [http://www.ocrwm.doe.gov/factsheets/doeymp0416.shtml]

OCRWM, "Switzerland's Radioactive Waste Management Program," Office of Civilian Radioactive Waste Management, U.S. Department of Energy, DOE/YMP-0417, June 2001 [http://www.ocrwm.doe.gov/factsheets/doeymp0417.shtml]

OCRWM, "The United Kingdom's Radioactive Waste Management Program," Office of Civilian Radioactive Waste Management, U.S. Department of Energy, DOE/YMP-0418, June 2001 [http://www.ocrwm.doe.gov/factsheets/doeymp0418.shtml]

OCRWM, "The U.S. Department of Energy's Role in the Nuclear Regulatory Commission's Licensing Process for a Repository," Office of Civilian Radioactive Waste Management, U.S. Department of Energy [http://www.ocrwm.doe.gov/factsheets/doeymp0113.shtml]

OCRWM, "Transportation of Spent Nuclear Fuel and High-Level Radioactive Waste to Yucca Mountain: Frequently Asked Questions," Office of Civilian Radioactive Waste Management, U.S. Department of Energy, DOE/ONT-0614, Rev. 1, January 2006 [http://www.ocrwm.doe.gov/transport/pdf/snf_transfaqs.pdf]

OCRWM, "Transporting Spent Nuclear Fuel and High-level Radioactive Waste to a National Repository: Answers to Frequently Asked Questions," Office of Civilian Radioactive Waste Management, U.S. Department of Energy [http://www.ocrwm.doe.gov/wat/pdf/snf_transfaqs.pdf]

OCRWM, 2007 "Repository Sites Considered in the United States," Yucca Mountain Project, Office of Civilian Radioactive Waste Management, U.S. Department of Energy [http://www.ocrwm.doe.gov/ym_repository/about_project/sitesconsidered.shtml]

OMB, "Budget of the United States–FY 2005," Office of Management and Budget, [http://www.whitehouse.gov/omb/budget/fy2005/pdf/budget/energy.pdf]

OMB, "Budget of the United States – FY 2006," Office of Management and Budget, [http://www.whitehouse.gov/omb/budget/fy2006/pdf/budget/energy.pdf]

ONEST, "Nuclear Energy Timeline and History of the Uranium Program," Office of Nuclear Energy, Science and Technology, U.S. Department of Energy, n.d. [http://www.ne.doe.gov/]

Ouchida, Kurt. "The Yucca Mountain Saga: A Project in Decay," [http:www.astro.com/yucca/html]

Pasternak, Douglas. "A nuclear nightmare," *U.S. News and World Report*, 17 September 2001, p. 44.

Pasternak, Douglas. "Assessing threats," *U.S. News and World Report*, 28 April 2003. p. 47.

PFS v. Utah, Case No. 2:01-CV-270C, 30 July 2002 [http://web.lexis-nexis.com/universe/printdoc]

PFS, "Atomic Safety and Licensing Board Recommends License for Spent Nuclear Site: First Such Recommendation in Nearly a Decade," News Release 24 February 2005, Private Fuel Storage, LLC [http://www.privatefuelstorage.com/whatsnew/newsreleases/nr2-24-05.html]

PFS, "FAQs: Transportation," Private Fuel Storage, LLC [http://www.privatefuelstorage.com/faqs/faq/faq-transportation.html]

PFS, "Licensing Board Denies on a Technicality PFS's 'Smaller Site' Proposal," Private Fuel Storage, LLC, [http://www.privatefuelstorage.com/whatsnew/whatsnew.html]

PFS, "Motions Filed in Skull Valley Band/Private Fuel Storage vs. State of Utah," News Release 12 December 2001, Private Fuel Storage, LLC [http://www.privatefuelstorage.com/whatsnew/newsreleases/nr12-12-01.html]

PFS, "Private Fuel Storage Challenges Licensing Board Decision," Private Fuel Storage, News Release 31 March 2003 [http://www.privatefuelstorage.com/whatsnew/newsreleases/nr3-31-03.html]

PFS, "Private Fuel Storage Wins Approval from Nuclear Regulatory Commission," News Release 9 September 2005, Private Fuel Storage [http://www.privatefuelstorage.com/whatsnew/newsreleases/nr9-09-05.html]

PFS, "The Licensing Process," Private Fuel Storage, LLC [http://www.privatefuelstorage.com/project/licensing.html]

PFS, "The PFS Facility," Private Fuel Storage, LLC [http://www.privatefuelstorage.com/project/facility.html]

PHMSA, "About PHMSA," Pipeline and Hazardous Materials Safety Administration, U.S. Department of Transportation [http://www.phmsa.dot.gov/about/index.html]

PHMSA, "Hazardous Materials Safety," Office of Hazardous Materials Safety, Pipeline and Hazardous Materials Safety Administration, U.S. Department of Transportation [http://www.phmsa.dot.gov/riskmgmt/risk.htm]

PL109-58, "Energy Policy Act of 2005," Public Law 109-58, 8 August 2005 [http://frwebgate.access.gpo.gov/cgi-bin/getdoc.cgi?dbname=109_cong_public_laws&docid=f:publ058.109.pdf]

POGO, "Nuclear Power Plant Security: Voices from Inside the Fences," 12 September 2002, Project on Government Oversight [http://www.POGO.org/p/environment/eo-020901-nukepower.html]

Public Citizen, "Another Nuclear Rip-off: Unmasking Private Fuel Storage," Public Citizen's Critical Mass Energy & Environment Program, July 2001 [http://www.citizen.org/documents/pfsreport.PDF]

Public Citizen, "A Brief History of Irradiated Nuclear Fuel Shipments: Atomic Waste Transport 'Incidents' and Accidents the Nuclear Power Industry Doesn't

Want You to Know About," Nuclear Information and Resource Service [http://www.citizen.org/print_article.cfm?ID= 5873]

Pulsiper, Allan G. "The Risk of Interim Storage of High Level Nuclear Waste," *Energy Policy*, July 1993: 798-812.

Raddatz, M.G. and M.D. Waters. "Information Handbook on Independent Spent Fuel Storage Installations," Spent Fuel Project Office, Office of Nuclear Material Safety and Safeguards, U.S. Nuclear Regulatory Commission, NUREG-1571, December 1996.

Reed, James B. "The State Role in Spent Fuel Transportation Safety: Year 2000 Update," NCSL Transportation Series, January 2000, No. 14 [http://www.ncsl.org/programs/transportation/transer14.htm]

Rissmiller, Kent. "Equality of Status, Inequality of Result: State Power and High-Level Radioactive Waste," *Publius: The Journal of Federalism*, Winter 1993: 103-18.

Roche, Lisa Riley. "Judge won't rule on wisdom of N-waste facility," *Deseret News*, 12 April 2002 [http://www.energy-net.org/IS/EN/NUKE/WST/FED/NEWS/NP-1212.402]

Rogers, Kenneth A. and Marvin G. Kingsley. "The Politics of Interim Radioactive Waste Storage: The United States," *Environmental Politics*, vol. 13, No. 3, September 2004: 590-611.

Rogers, Kenneth A. and Marvin G. Kingsley. "The Transportation of Highly Radioactive Waste: Implications for Homeland Security," *Journal of Homeland Security and Emergency Management*, Volume 1, Issue 2, Article 13, 2004.

Rogers, Kenneth A. and Donna L. Rogers. "The Politics of Long-Term Highly Radioactive Waste Disposal," *Midsouth Political Science Review*, vol. 2, 1998: 61-71.

Rosenbaum, David B. "Yucca Mountain Draws Fire Even From Some within DOE," *Engineering News Report*, 27 April 1992: 10.

Ryan, Cy. "Nuclear dump in Utah set for approval," *Las Vegas SUN*, 10 February 2003 [http://www.lasvegassun.com/sunbin/stories/text/2003/feb/10/514642537.html]

Shapiro, Fred C. "Yucca Mountain," *The New Yorker*, May 22, 1988, in *Taking Sides: Clashing Views on Controversial Issues*, 3rd. ed., Theodore D. Goldfarb, ed. (Dushkin Publishing Group, Inc.: Guilford, CT, 1989), pp. 260-69.

Spangler, Jerry D. and Lee Davidson. "Utilities' promise to Bennett, Hatch may carry little weight," *Deseret News*, 14 July 2002 [http://deseretnews.com/dn/view/0,1249,405017719,00.html]

Sprung, J.L., D.J. Ammerman, J.A. Koski, and R.F. Weiner. "Spent Nuclear Fuel Transportation Package Performance Study Issues Report," Sandia National Laboratories, NUREG/CR-6768, SAND2001-0821P, 31 January 2001 [http://www.nrc.gov/reading-rm/doc-collections/nuregs/contract/cr6768/nureg-6768.pdf]

Stanton, Lawrence. Deputy Director, Protective Security Division, Information Analysis and Infrastructure Protection, U.S. Department of Homeland Security, Interview, Washington D.C., 20 May 2005.

State of Nevada, "Concerns About Yucca Mountain," State of Nevada Nuclear Waste Project Office [http://www.well.com/user/rscime/yucca/concerns.html]

State of Nevada, "State of Nevada Socioeconomic Studies: Biannual Report, 1993-1995," State of Nevada Nuclear Waste Project Office [http://www.state.nv.us/nucwaste/yucca/sebian95.html]

State of Nevada, "Why Nevada is Opposed to Yucca Mountain," State of Nevada Nuclear Waste Project Office [http://www.well.com/user/rscime/yucca/wyop1.html]

Struglinski, Suzanne. "NRC: Casks would survive big blaze," *Las Vegas SUN*, 15 September 2005 [http://www.lasvegassun.com/sunbin/stories/sun/2005/sep/15/519360447.html]

Struglinski, Suzanne. "Nuclear transport to Utah may face problems," *Las Vegas Sun*, 12 September 2005 [http://www.lasvegassun.com/sunbin/stories/sun/2005/sep/12/519342580.html?struglinski]

Struglinski, Suzanne. "Nuclear waste site looks doomed," *deseretnews.com,* 8 September 2006 [http:deseretnews.com/dn/print/1,1442,645199773,00.html]

Struglinski, Suzanne. "PFS is still optimistic: Firm's chief aims to get interim storage in Utah," *deseretnews.com,* 28 September 2006 [http:deseretnews.com/dn/print/0,1249,650194421,00.html]

Stuart, Ivan F. "Owl Creek Energy Project: A Solution to the Spent Fuel Temporary Storage Issue," Waste Management Conference, 28 February-4 March 1999.

STUK, *Joint Convention on the Safety of Spent Fuel Management and on the Safety of Radioactive Waste Management*, 2nd Finnish National Report as referred to in Article 32 of the Convention, STUK-B-YTO 243, October 2005 [http://www.stuk.fi/julkaisut/stuk-b/stuk-b-yto243.pdf]

Suplee, Curt. "A Nuclear Problem Keeps Growing," *The Washington Post*, 31 December 1995: A1, A18, A19.

Talhelm, Jennifer. "Skull Valley waste site blocked, at least for now," *Daily Herald*, 16 September 2005 [http://www.harktheherald.com/modules.php?op=modload&name=News&file=article&sid=64418]

TEC/WG, "TEC Members," Transportation External Working Group, U.S. Department of Energy, 2007 [http://www.tecworkinggroup.org/members.html]

TEC/WG, "The Transportation External Coordination Working Group," National Transportation Program, U.S. Department of Energy, Revised April 2001 [http://www.ntp.doe.gov/ftplink/fec_wg.pdf]

TEC/WG, "Transportation External Coordination Working Group (TEC) 25-27 July 2000 Meeting Summary," Transportation External Coordination Group, 26 July 2000 [http://www.ntp.doe.gov/tec/ind2000.pdf]

TEPP, "Transportation Emergency Preparedness Program," U.S. Department of Energy, Office of Environmental Management, Office of Transportation, February 2002 [http://www.em.doe.gov/otem/TEPPfactsheet02-20-02.pdf]

Tetreault, Steve. "Congress settles on budget for Yucca Mountain," Las Vegas Review-Journal, 6 November 2003. [http://www.reviewjournal.com/lvrj_home/2003/Nov-06-Thu-2003/news/22528355.html]

Tetreault, Steve. "Nuclear Waste Shipments: Report: Department not fully prepared," *Las Vegas Review-Journal*, 9 February 2002 [http://www.energy-net.org/IS/EN/NUKE/WST/FED/NEWS/NP-099.202]

Tetreault, Steve. "Nye County seeks role in nuclear waste project," Las Vegas Review-Journal, February 9, 2002. Available at [http://www.reviewjournal.com]

Travers, William D. "Fitness-for-Duty Enhancements to Address Concerns Regarding Fatigue of Nuclear Facility Security Force Personnel–Draft Order and Compensatory Measures for Power Reactor Licensees," Executive Director for Operations, Nuclear Regulatory Commission, 10 March 2003 (Memorandum COMSECY-03-0012, Attachment 3)

T-REX, "A Consideration of Routes and Modes in the Transportation of Radioactive Materials and Waste: An Annotated Bibliography," Transportation Resource Exchange Center, ATR Institute, University of New Mexico, 31 December 2000 [http://www.trex-center.org/dox/routesbib.pdf]

T-REX, "Spent Nuclear Fuel Transportation Cask Tests," Transportation Resource Exchange Center (T-REX) [http://www.trex-center.org/casktest.asp]

T-REX, "Spent Nuclear Fuel Transportation Regulations," Transportation Resource Exchange Center (T-REX) [http://www.trex-center.org/caskregs.asp]

T-REX, "Transportation Resource Exchange Center," Transportation Resource Exchange Center (T-REX) [http://www.trex-center.org]

trib.com, "Company still pushing nuclear dump," *Casper Star-Tribune*, Tuesday, 28 August 2001 [http://www.state.nv.us/nucwaste/news2001/nn11383.htm]

URL, "Underground Research Laboratory," Applied Seismology Laboratory, Liverpool University Department of Earth Sciences, 2006 [http://www.liv.ac.uk/seismic/research/url/url.html]

U.S. Congress, "Final Vote Results for Roll Call 133," H.J. 87 [http://clerkweb.house.gov/cgi-bin/vote.exe?year=2002&rollnumber=133]

U.S. Congress, "U.S. Senate Roll Call Votes," S.J. 34, 107th Congress, 2nd Session [http://www.senate.gov/legislative/vote1072/vote_00167.html]

U.S. Congress, H.R. 1270, House, 105th Congress, 1st Session, Bill Summary and Status, Vote 557, 30 October 1997 [http://thomas.loc.gov/cgi-bin/bdquery/z?d105:HR01270:@@@X]

U.S. Congress, H.R. 1270, Senate, 105th Congress, 2nd Session, Vote 148, 2 June 1998 [http://www.senate.gov/legislative/vote1052/vote_00148.html]

U.S. Congress, S. 104, Senate, "Nuclear Waste Policy Act of 1997," 105th Congress, 1st Session, [http://thomas.loc.gov/cgi-bin/bdquery/z?d105:SN00104:@@@L&summ2=m&]

U.S. Congress, S. 1271, "Nuclear Waste Policy Act of 1996," *Congressional Record*, 104th Congress, 2d Session.

U.S. Congress, S. 1287, Senate, "Nuclear Waste Policy Amendments Act of 2000," 106th Congress, 1st Session [http://thomas.loc.gov/cgi-bin/bdquerytr/z?d106:SN01287:@@@D&summ2=m&]

U.S. Congress, S. 1287, Senate, 106th Congress, 2nd Session, Vote 88, 2 May 2000 [http://thomas.loc.gov/cgi-bin/bdquery/z?d106:SN01287:@@@D]

U.S. Congress, S. 14, Senate, "A Bill to Enhance Energy Security of the United States," 108th Congress, 2nd Session [http://thomas.loc.gov/cgi-bin/bdquery/z?d108:s.00014:]

U.S. Congress, S. 1936, "Nuclear Waste Policy Act of 1996," *Congressional Record*, 104th Congress, 2d Session.

USCM, "Resolution: Transportation of High-Level Nuclear Waste," U.S. Conference of Mayors, Adopted at the 70th Annual Conference of Mayors, Madison, WI, 14-18 June 2002.

Utah, DEQ, "State of Utah's Contentions on the Construction and Operating License Application by Private Fuel Storage", LLC for an Independent Spent Fuel Storage Facility," U.S. Nuclear Regulatory Commission, Docket No. 72-22-ISFSI, 23 November 1997. Available at: [http://www.deq.utah.gov/Issues/no_high_level_waste/documents/11_23_97.PDF]

Utah DEQ, "Comments Submitted by the State of Utah on the Draft Environmental Impact Statement (DEIS)," Utah Department of Environmental Quality, DEIS Comments Index, 20 September 2000 [http://www.deq.state.ut.us/EQOAS/no_high_level_waste/commentsindex.htm]

Utah DEQ, "Information on Private Fuel Storage's Proposal to Locate a High Level Nuclear Waste Storage Facility on the Skull Valley Goshute Indian Reservation," State Concerns Statement, Utah Department of Environmental Quality [http://www.deq.state.ut.us/EQOAS/no_high_level_waste/concerns.htm]

Utah v. PFS, 2000) U.S. Court of Appeals for the Tenth Circuit, No. 99-4104, 25 April 2000. [http://web.lexis-nexis.com/]

Utah.gov, "Governor announces opposition to proposed high level nuclear waste dump in Utah," 14 April 1997 [http://yeehaw.state.ut.us/]

Utah.gov, "Governor Signs Legislation Blocking High-Level Nuclear Waste From Coming to Utah," 13 March 2001 [http://yeehaw.state.ut.us/]

WGA, "Assessing the Risks of Terrorism and Sabotage Against High-Level Nuclear Waste Shipments to a Geologic Repository or Interim Storage Facility," Western Governors' Association, Policy Resolution 01-03, August 14, 2001 [http://www.westgov.org/wga/policy/01/01_03.pdf] Originally adopted as Policy Resolution 98-008.

WGA, "Policy Resolution 00-031–Private Storage of Commercial Spent Nuclear Fuel," Western Governors' Association, Policy Resolution 00-031, 13 June 2000.

WGA, "Transportation of Spent Nuclear Fuel and High-Level Radioactive Waste," Western Governors' Association, Policy Resolution 02-05, June 25, 2002 [http://www.westgov.org/wga/policy/02/nuketrans_05.pdf]

White House, "National Energy Policy," Report of the National Energy Development Group, May 2001 [http://www.whitehouse.gov/energy/Forward.pdf]

White House, "National Energy Policy," White House, 2001 [http://www.whitehouse.gov/news/releases/2001/05/20010517-2.html]

WIEB, "HLW Committee Comments on DOE's April 30, 1998 Notice of Revised Policy and Procedures for Safe Transportation and Emergency Response Training; Technical Assistance and Funding for Spent Nuclear Fuel and High-Level Radioactive Waste (Notice)," Western Interstate Energy Board, 31 July 1998 [http://www.westgov.org/wieb/reports/180c1998.htm]

WIEB, "HLW Committee Comments on DOE's Notice of Proposed Policy and Procedures for Safe Transportation and Emergency Response Training for Spent Nuclear Fuel and High-Level Radioactive Waste (Notice)," Western Interstate

Energy Board, 12 September 1996 [http://www.westgov.org/wieb/reports/180c1998.htm]

WIEB, "Reports and Comments," Western Interstate Energy Board [http://www.westgov.org/wieb/reports.html]

Witt, James Lee. *Stronger in the Broken Places* (Times Books, NY: 2002).

Witt, James Lee. "Review of Emergency Preparedness of Areas Adjacent to Indian Point and Millstone," James Lee Witt Associates, LLC, 7 March 2003 [http://wittassociates.com/news_final_report_NY.html]

Zacha, Nancy J. "Fear of Shipping," Editor's Note, *Radwaste Solutions*, March/April 2003 [http://www2.ans.org/pubs/magazines/rs/pdfs/2003-3-4-2.pdf]

Index